Pest Management and Phytosanitary Trade Barriers

Pest Management and Phytosanitary Trade Barriers

Neil W. Heather

School of Land, Crop and Food Sciences, University of Queensland, Gatton, Queensland, Australia

and

Guy J. Hallman

United States Department of Agriculture, Agricultural Research Service, Weslaco, Texas, USA

www.cabi.org

CABI is a trading name of CAB International

CABI Head Office
Nosworthy Way
Wallingford
Oxfordshire OX10 8DE
UK
Tel: +44 (0)1491 832111

E-mail: cabi@cabi.org
Website: www.cabi.org

CABI North American Office
875 Massachusetts Avenue
7th Floor
Cambridge, MA 02139
USA
Tel: +1 617 395 4056

E-mail: cabi-nao@cabi.org

A catalogue record for this book is available from the British Library, London, UK.

Library of Congress Cataloging-in-Publication Data

Heather, Neil W.
Pest management and phytosanitary trade barriers / Neil W. Heather and Guy J. Hallman.
p. cm.
Includes bibliographical references and index.
ISBN 978-1-84593-343-2 (alk. paper)
1. Food crops--Protection. 2. Food crops--Losses--Prevention. 3. Food crops--Postharvest losses--Prevention. 4. Insect pests--Control. 5. Bioterrorism. I. Hallman, Guy J. II. Title.

SB950.H38 2008
632'.7--dc22 2007026971

ISBN: 978 1 84593 343 2

Typeset by Columns Design Ltd, Reading, UK
Printed and bound in the UK by CPI Group (UK) Ltd, Croydon, CR0 4YY, UK.

First published in 2008. Transferred to POD in 2020.

Contents

Preface

The origin of the scope of this volume lies with the experiences of the authors as researchers and has been predominantly on postharvest phytosanitary problems involving insect pests. A large part of this experience has been the wisdom imparted by research and regulatory colleagues throughout the world. We are particularly indebted to colleagues in organizations such as the Food and Agriculture Organization of the United Nations (FAO), the International Atomic Energy Agency (IAEA), the United States Department of Agriculture (USDA) and Australian federal and state departments of agriculture and universities.

Agricultural produce is a basic need of humanity and its production has become increasingly internationalized over the past two centuries. This, together with developments in transportation, has led to a burgeoning of trade and with it a need for producing countries to ensure as far as possible that they do not acquire pests which they do not already possess. The answer to this lies in a large measure with phytosanitary practices and strategies which permit the transfer of produce but filter out associated unwanted pests.

Our book, essentially entomological, is intended to supplement the experience of researchers whose role it is to develop pest management strategies and to guide pest regulatory workers in the selection of requirements to achieve phytosanitary security with minimal disruption to trade including unacceptable injury to fragile commodities. We are also hopeful that educators will see the need for formal training in phytosanitation and that this volume might make a contribution.

Fortunately the publisher CABI is aware of the value of dissemination of knowledge borne of experience before it becomes lost. Their editorial expertise and long association with agricultural research has assisted greatly in the formulation of this volume. Finally, we are greatly indebted to colleagues, acknowledged separately, who generously reviewed chapters for us.

Neil W. Heather and Guy J. Hallman

Acknowledgements

We wish to acknowledge the organizational support past and present which has enabled this book to be compiled. For Guy Hallman, it is the Agricultural Research Service (ARS) of the United States Department of Agriculture (USDA) and for Neil Heather, in the past it was the Queensland Department of Primary Industries Entomology Branch and subsequently the Integrated Pest Management Group of the University of Queensland Gatton, Faculty of Natural Resources and Veterinary Science. Technical chapters have been generously reviewed or material provided to assist in their preparation by the colleagues listed below. They enabled us to remove many incipient errors and inconsistencies but final responsibility for the text is ours alone as authors.

Chapter 2: Mark Wheelis, University of California at Davis, USA; Victoria Yokoyama, ARS, Parlier, California, USA; and Clifford Ramos, California Department of Food and Agriculture, Sacramento, California, USA (information on Mediterranean fruit fly).

Chapter 3: Robert Griffin, Animal and Plant Health Inspection Service (APHIS), USDA, Raleigh, North Carolina, USA; Chris W. Hood, Australian Government Department of Agriculture, Fisheries and Forestry, Canberra, Australia; and Timothy Lockley, APHIS, Gulfport, Mississippi, USA (permission to use cartoon).

Chapter 4: Robert Sutherst, University of Queensland, St Lucia, Queensland, Australia (formerly Division of Entomology, Commonwealth Scientific and Industrial Research Organization (CSIRO), Indooroopilly, Queensland).

Chapter 5: Annice C. Lloyd, Department of Primary Industries, Indooroopilly, Queensland, Australia; and Rosemary Kopittke, Department of Primary Industries, Indooroopilly, Queensland, Australia.

Chapter 6: Rosemary Kopittke, Department of Primary Industries, Indooroopilly, Queensland, Australia; Bruce Champ, formerly Australian Centre for International Agricultural Research, Canberra, Australian Capital Territory (ACT), Australia;

Christina Devorshak, APHIS, Raleigh, North Carolina, USA; and Pauline Wyatt, Department of Primary Industries, Indooroopilly, Queensland, Australia.

Chapter 7: Walter P. Gould, APHIS, USDA, Riverdale, Maryland, USA (and contribution of a personal observation); and Anthony B. Ware, Citrus Research International, Nelspruit, South Africa.

Chapter 8: Donald Thomas, ARS, Weslaco, Texas, USA; and Robert Corcoran, formerly Department of Primary Industries, Cairns, Queensland, Australia.

Chapter 9: John R. Logar, Sterigenics International, Swedesboro, New Jersey, USA; and Nichole M. Levang-Brilz, APHIS, Raleigh, North Carolina, USA.

Chapter 10: Robert Winks, formerly Stored Grains Research Laboratory, CSIRO, Canberra, ACT, Australia; and James Leesch, ARS, Parlier, California, USA.

Chapter 11: J. Larry Zettler, APHIS, Raleigh, North Carolina, USA.

Chapter 12: Robert Corcoran, formerly Department of Primary Industries, Cairns, Queensland, Australia; and Errol Hassan, University of Queensland, Gatton, Queensland, Australia.

Chapter 13: Kevin V. Donohue, Department of Entomology, North Carolina State University, Raleigh, North Carolina, USA; and James Hansen, formerly ARS, Yakima, Washington, USA.

In addition to reviewers, thanks are due to colleagues who assisted with procurement and presentation of illustrations. In particular, we acknowledge Allen Bruce, formerly of the Integrated Pest Management Group, University of Queensland, Gatton and Maryanne Kepui, Australian Quarantine Inspection Service, Brisbane. Photographers are acknowledged in the captions of illustrations with the exception of those by the authors.

1 Pest Management and Phytosanitary Trade Barriers

The agricultural species that form the basis of the economies of most of the countries of the world have been largely introduced from other areas, clearly demonstrating that importation of alien species is not inherently detrimental at least from the human standpoint. However, many pests have been transported around the globe as well and cause a great amount of damage. It is estimated that the loss caused by invasive species globally is about US$1.4 $\times$ 10^{12} or 5% of the world gross national product (Pimentel *et al.*, 2007). Although tens of thousands of species have invaded other lands, millions have not. Undoubtedly a great many potentially invasive species could cause significant economic and ecological damage to the diverse countries of the world. Phytosanitation aims to keep that damage and the number of new invasive species as low as possible through regulation of trade of items that could carry invasive species. But these requirements are a primary impediment to international trade, a key and growing component of most economies (Fig. 1.1). Mumford (2002) points out that domestic consumers ultimately pay for quarantine restrictions in higher prices for quarantined goods while domestic producers of those goods or, we might add, reasonable replacements for them, benefit.

Reasons for phytosanitary barriers to trade can be variable:

- Quarantine against known pests.
- Unknown risks of invasive species.
- Political and trade drivers.
- Financial costs.
- Social costs.

In conformity with the World Trade Organization (WTO) Sanitary and Phytosanitary (SPS) Agreement (WTO, 2007a) categorized quarantine pests are restricted to those that are absent from the place which takes action to prohibit entry of the pest or its host if this could lead to entry and establishment, or to pests which are present but suppressed or contained. This latter category can be

Fig. 1.1. Section of the large Tokyo fresh fruit and vegetable market, Ohta Ichiba, typical of major metropolitan markets supplied in large part by internationally sourced produce subject to phytosanitary requirements (Source: reproduced courtesy of the Australian Institute of Agricultural Science and Technology).

termed 'regulated non-quarantine' pests if they are the subjects of relevant regulatory proclamations. The modern approach is to apply phytosanitary quarantine restrictions against all imported commodities unless there is agreement otherwise. This contrasts with the original use of quarantine, which would have been directed at a particular commodity or pest.

Most regulated non-quarantine pests are cosmopolitan pests that reached their worldwide dispersal potential before the implementation of modern quarantine. Although these cosmopolitan pests are not categorized as quarantine pests they can be subjected to phytosanitary regulation for a number of reasons, the most common being to manage pest populations in areas where the species are at a low level of incidence because of the implementation of a system approach (Chapter 5).

Regulated Pests and their Origins

The historic Silk Road system of trade routes across Asia would have been the origin of much of the early human initiated dispersal of many, now cosmopolitan, oriental pest species to Europe and the converse, together with the dispersal of pests from and to places en route. The pests most amenable to this type of movement were those associated with durable commodities and staple diet consumables. However, a degree of susceptibility to dispersal is evident, with some grain and structural fabric pests proving highly mobile and now cosmopolitan while others are still relatively restricted to their original areas. These latter, such as the very destructive dermestid khapra beetle, *Trogoderma granarium*, and the bostrychid larger grain borer, *Prostephanus truncatus*, have achieved major status as quarantine pests in recent times because of their still limited distribution. Nevertheless most pests of durable commodities had achieved cosmopolitan status before the advent of modern phytosanitary practice and these pests have become even more widely dispersed through recent sea and air trade.

As trading ships became faster the type of host commodity transported became more diverse and with it the pests associated with more perishable commodities as these were included in trade or carried as sustenance of ships' crews. Dispersal of the Mediterranean fruit fly, *Ceratitis capitata*, illustrates this well (Maddison and Bartlett, 1989). A major host is citrus and early voyagers learned the benefits of fresh citrus to prevent scurvy. Citrus was cultivated in the Asian subcontinent from earliest times (Willis, 1966) but with the advent of trade became widely cultivated in the Mediterranean region especially the Iberian and Italian peninsulas where Mediterranean fruit fly was probably endemic by then, despite its southern African region origins. With the development of sea trade, convenient ports were developed for watering, fuelling and replenishment of food including fresh fruit and vegetables. Today many of these can be identified as areas of establishment of Mediterranean fruit fly including the Canary Islands, St Helena, Cape Province South Africa, south-west Western Australia, Hawaii, Central America and landfalls in South America such as Rio de Janeiro. Undoubtedly there were other areas where the species failed to establish initially or failed to survive long term including eastern Australia and New Zealand.

A source of confirmation of pest dispersal in this way can often be found in the label data on specimens in entomological reference collections in the locations in question and elsewhere the species might have been of interest. Many are recorded in distribution data of taxonomic papers on the pest species. Care must be exercised to differentiate between specimens taken as interceptions at entry and those from established populations at the recorded location. However, even interception records are valuable in that they indicate the possibility of establishment on that or other occasions. The outcome of this historical process is that many pests will be found to have reached their limit of dispersal before phytosanitary quarantine became an established practice. In some places they will be recognizably endemic and consequently of no justifiable quarantine significance. In other places where establishment potential is marginal they may be present and persisting below the limit of ordinary detection. If this can be determined, there might be no justification for quarantine barriers to trade with respect to that pest. The reliability of pest incidence data is in direct relation to the search effort put into detection surveys.

Modern Phytosanitary Practice

Phytosanitary measures relate to quarantine or non-quarantine pests, legislatively categorized, nominated or gazetted. Ebbels (2003) provided a detailed coverage of the history of phytosanitation. While the origins of legislated controls against important pests and diseases are to be found in Europe, dating from 1660, these were ad hoc measures aimed at specific pests and diseases. Holistic, legislatively enforceable plant quarantine dates from the US Plant Quarantine Act of 1912. This approach came to be adopted widely throughout the world.

Phytosanitary certification

Although phytosanitary certification pre-dates the International Plant Protection Convention (IPPC), the IPPC has developed and promoted an international phytosanitary pro forma (see Appendix I) with the following requirements (FAO, 1997).

Each contracting government shall make arrangements for the issue of phytosanitary certificates to accord with the plant protection regulations of other contracting governments, and in conformity with the following provisions:

- Inspection shall be carried out and certificate issued only by or under the authority of technically qualified and duly authorized officers and in such circumstances and with such knowledge and information available to those officers that the authorities of importing countries may accept such certificates with confidence as dependable documents.
- Each certificate covering materials intended for planting or propagation shall be as worded in the Annex to this Convention and shall include such additional declarations as may be required by the importing country. The model certificate may also be used for other appropriate plants or plant products where it is not inconsistent with the requirements of the importing country.
- The certificates shall bear no alterations or erasures.

Each contracting government undertakes to ensure that consignments of plants intended for planting or propagation that are imported into its territories are accompanied by certificates consistent with the model set out in the Annex to the Convention.

Role of WTO

The WTO has evolved from the General Agreement on Tariffs and Trade (GATT) and dates from 1947 when it was established as an international forum to encourage free trade between member states, currently numbering 151. One of the outcomes of the 1986–1994 Uruguay Round of negotiations of GATT was the Agreement on Application of Sanitary and Phytosanitary Measures, which included a requirement that quarantine restrictions 'must have a scientific basis'. This became known as the 'SPS Agreement' and has had a major influence on the exercise of phytosanitary constraints on trade (WTO, 2007a). The use of international standards or codes of practice to achieve harmonization through a scientific basis of quarantine requirements was recommended and their development is ongoing.

General considerations

Trade and political considerations can drive biological barriers to trade. Protection of existing or future production industries from new pests is paramount. Subject to the agreement of a country to conform with the principles inherent in the SPS Agreement there will be political and industry pressures on regulatory organizations to adopt an extreme conservative approach to pest risk where it is otherwise advantageous (Roberts and Krissoff, 2004). The SPS Agreement requires sound scientific justification of any phytosanitary trade barrier but recognizes the ultimate sovereignty of a country, giving it the right to impose import prohibitions. However, since trade is a two-way process, on balance it is normally preferable for a country to adopt a position that permits trade to proceed with reasonable levels of risk management. This requires a balanced public information approach.

Environmental considerations have become increasingly important in recent times (Hallman, 2007). Apart from global climate change, discussed in detail in Chapter 4 the risk to native and introduced ornamental vegetation from pests not previously present can affect public quality of life. This can have a financial flow-on to tourism industries and to costs of control of the new pest. Eradication or biological control programmes may be called for and, while these are usually the most cost effective, in the long term initial costs are invariably high.

The financial and social impacts of phytosanitary action are likely to be complex, being both positive and negative. First, they can usually be justified on grounds of avoidance of pest management costs that would be experienced if a destructive pest gained entry to a country from which it was previously absent. Secondly, there is the avoidance of potential costs of eradication as an alternative to ongoing pest management. Where an industry relies on freedom from pests as a condition of entry to another market there is a potential loss of profits should the pest in question gain entry. This can have social ramifications especially if the industry is labour intensive, depriving individuals of an income that permits a particular way of life. But the actions can have a broader social impact where a population is denied access to a staple commodity at a price which would be much less if imports were permitted, both through actual costs and absence of competition.

The impact of phytosanitary measures on trade needs to be managed if trade is to be permitted to flow. Routine pest management measures can be involved and new ones developed that enable quarantine restrictions to function as a filter for unwanted pest organisms. These measures can come from a wide variety of sources covering both pre- and postharvest phases of crop production. In this volume we have sought to identify these sources, methods of pest management and governing factors and to present them in a way that facilitates formulation of the best combination of measures for a particular trading situation.

Phytosanitation has developed with increasing momentum as an important component of pest management in the production and marketing of agricultural produce over the past three centuries. Since antiquity, man has increased the natural dispersal of commercial crop plants and others through migration, trade and travel. The dispersal patterns of many associated plant pests mirror traditional travel and trade routes on land and water. Modern transportation has added air transport to this pattern. While many pests, plant pathogens and pest plants are now cosmopolitan in their distribution others have yet to reach their full dispersal potential for one reason or another. Phytosanitary measures can be a cost effective component of holistic pest management in the latter case. Since pest management has political and social as well as economic and volume production drivers governmental regulation has become increasingly involved. Resultant phytosanitary requirements affect the flow of trade and will continue to do so.

Phytosanitary barriers cannot offer absolute security against the entry of a pest but risk management will reduce the possibility of entry and establishment to acceptable levels. It could be argued that only prohibition of trade can confer absolute security and it is the only option in some circumstances. However, if there is a demand for a commodity and it is not available from pest free sources this increases the exposure of the importing area to contraband which is likely to be a much greater risk than the efficacy achievable in a managed system. Therefore, the process of phytosanitary quarantine necessarily becomes one of risk management if trade is to proceed.

Phytosanitary Factors Governing Trade

Trade patterns relative to pest species distribution and commodity production locations are the main drivers of phytosanitary action. However, from country to country, pests will vary in their perceived risk to the agricultural industries and to the environment.

The International Plant Protection Convention of the United Nations Food and Agriculture Organization (FAO, 1997) gives eight basic principles applicable to phytosanitary import and export actions. These are:

- Sovereignty, which recognizes the right of a country to safeguard and manage its production of agricultural commodities.
- Necessity, which can arise from a need to react initially to acute problems or emergencies before full information is available.
- Minimal impact, which requires that phytosanitary action be no more than is adequate to control a problem.
- Transparency, in that reasons for phytosanitary actions be made known.
- Modification, which means that flexibility should be inherent in the response to phytosanitary problems so that developing knowledge can be incorporated readily into management of the problem.
- Harmonization, which accepts that measures and procedures to manage problems in one part of the world and which are shown to be effective, should be accepted elsewhere.
- Equivalence recognizes that procedures effective against a pest can vary from country to country without significant detriment to efficacy.
- Dispute settlement, which involves a willingness by countries to submit to adjudication of technical and other disputes by an independent body, for example the WTO.

In addition, the IPPC give a further eight specific principles, viz.:

- Cooperation.
- Technical authority.
- Risk analysis.
- Managed risk.
- Pest free areas.
- Emergency action.
- Notification of non-compliance.
- Non-discrimination.

Extent of Trade in Commodities Susceptible to Phytosanitary Constraints

The value of modern world trade in agricultural, including horticultural, commodities for 2005 was US$852 billion (WTO, 2007b). This represented 8.4% of the value of all world trade and an increase of 8% over 2004. Information on the volumes and diversity of the plethora of agricultural commodities traded is not so readily available. Food constitutes 80% in value of agricultural commodities traded, the remainder being raw materials such as cotton, tobacco and leather. Both food and raw materials may be subject to phytosanitary constraints. But it does not stop there; actionable pests may be found in anything that crosses quarantine barriers. Major forest, agricultural, health and ecological threats, such as the Asian longhorn beetle, *Anoplophora glabripennis*, and the Argentine fire ant, *Solenopsis invicta*, were transmitted by wooden pallets, packing materials and conveyances not associated with agricultural trade. Quarantine pests and invasive species may enter any export container at virtually any point in the shipping train, especially flying insects attracted to lighting in warehouses and ports.

Phytosanitary problems reflect geographic and social patterns of trade. For each WTO-defined region, trade patterns are influenced by regional internal and external exporting and importing. Approximate values of agricultural produce have been estimated from 2005 WTO statistical data for all commodities as a guide (WTO, 2007b). In terms of major trading flows, for the European Union, 82% of agricultural exports in 2005, totalling US$370 billion in value, was with countries within the region and most of the remaining 18% was with North America and Asia. For the Asian region, which includes Australasia, the total was US$136 billion with 65% within the region, 18% with Europe and 17% with North America. For the North American trading region the value of trade was US$241 billion, with 42% within the region, 27% with Asia and the rest with other regions of the world. The Central and South American region (US$28 billion) traded mainly to Europe. The European Union region was the largest trader in 2005 with 43% of world agricultural commodities trade.

Major external exporters of agricultural produce in 2005 were North America with US$71 billion, the European Union with US$67 billion, and Brazil with US$35 billion (2%). Worldwide, the highest value importers were the European Union with US$97 billion, the USA with US$68 billion and Japan with

US$66 billion. Values give some indication of market distribution and consequently, established phytosanitary requirements that will need to be met by intending exporters to those markets.

Anticipation of phytosanitary constraints on emerging trade can be complex and needs to be assessed on a country-to-country basis, as phytosanitary measures are regulated by national entities despite the dependence of potential pest distribution on geographic and climatic parameters. Clearly, however, the largest markets are Europe, North America and in Asia, mainly Japan. The origins of their imports are likely to be diverse and each production area from which they are derived will almost certainly have a unique make-up of pest fauna. The IPPC (see Chapter 3) has as one of its roles a harmonizing effect on requirements but all importing countries retain national sovereignty over phytosanitary policies. International leadership in phytosanitary regulation lies with these national groupings with the USA having a major involvement with both imports and exports.

Rate of growth in trade

During 2005 the growth in world trade in agricultural commodities rose by 8%; it has varied in other recent years from 5 to 17%. Overall, long-term growth of both exports and imports can be expected in agricultural commodities in line with growth in all trade. For 2005, the European Union increase was 6–7% following 12–14% in the previous year. Other regions registered growth ranging from 19% for exports from China to 4 and 9% for the USA exports and imports, respectively. It can be expected that there will be continuing changes in linkage of growth and diversification of import and export countries and consequently upon the impact of phytosanitary measures on trade as newer importing countries seek to safeguard their own production industries.

Effects of free trade agreements

Free trade agreements can be expected to focus phytosanitation on technical issues and lead to eventual harmonization of requirements. Free trade agreements can have the effect of concentrating trade between countries or regions that may differ from overall world patterns. Given that the time required to negotiate and to develop phytosanitary treatments can take many years it is important for involved countries to anticipate free trade agreements in the planning of phytosanitary research programmes.

Pest Management in Relation to Phytosanitary Barriers

This book describes phytosanitary measures used to overcome phytosanitary barriers to trade. Trade problems and possible phytosanitary measures are summarized below.

Types and levels of barriers

Phytosanitary barriers range from absolute embargoes on trade to response measures that can be taken when pests are detected during surveillance at entry to a country. In general they can be:

- Categorization, nomination or gazettal of prohibited pests and associated commodities.
- Blanket prohibitions such that only commodities that are specifically approved may enter an area.
- Legislated prohibitions on the entry of certain commodities, supported by active regulatory surveillance.
- Legislated requirement for a commodity to have been subjected to a certified treatment previously demonstrated scientifically to achieve a level of efficacy against nominated pests agreed by the import regulator to be adequate to achieve a required level of quarantine security.
- Certified production and handling of the commodity under systems that will ensure freedom from nominated pests at a level adequate to meet the security judged to be required.
- Surveillance at selected points in the production, handling and shipping system usually by sampling and inspection and the refusal of entry, destruction or application of disinfestation measures at the discretion of the inspecting authority.

Scope of commodities experiencing problems

Virtually all agricultural and many non-agricultural commodities can be the subject of phytosanitary measures at some time or another because they are either hosts or incidental carriers of unwanted pest organisms. Although the tonnages of durable commodities in world trade, such as food grains, far outweigh perishable commodities it is the perishable commodities, mainly fruit, vegetables and cut flowers that are most intensively regulated for quarantine and other phytosanitary reasons. Durable commodities were more likely to have been traded over long distances in earlier times and hence their pest fauna became cosmopolitan to the extent of their biological potential before the development of surveillance services regulating quarantine and phytosanitation. Apart from being a more recent trade development, perishable commodities tend to have higher unit values and are more likely to carry potentially devastating production pests rather than pests of perishable commodities. However, durable commodities and their conveyances often harbour quarantine pests. Ceramic and marble tiles from Italy are one of the most heavily fumigated imports in the USA and other countries because of the occurrence of quarantined snails and other organisms.

Despite the cosmopolitan nature of most pests of durable commodities a few remain that are subject to wide-ranging, quarantine-based prohibitions. Of these, the aforementioned khapra beetle and the larger grain borer are possibly of greatest importance. Other pests of durables may be categorized as quarantine pests in more restricted areas. These include dermestids such as the warehouse beetle, *Trogoderma variabile*, that are limited in distribution in countries such as Australia. Pests of durables including the curculionid rice weevil, *Sitophilus oryzae*, have been subject to quarantine categorization in some areas such as parts of Europe based on their non-occurrence. But, distribution of such pests is largely governed by climatic suitability and the occurrence of infestations in traded goods does not represent a quarantine threat, as the species will not persist in areas where it does not currently exist. Quarantines have been justified in the past on the basis of pesticide resistance in populations in the area of origin. However, most large pest populations would have the genetic capacity, given time, to develop and maintain resistance if appropriate selection pressure exists.

Phytosanitary measures can be required as a trade strategy to give a quality advantage. Australia imposes quality standards on cereal and other grains including freedom from pests under an Act of Parliament with regulations supervised by the Australian Quarantine Inspection Service (Pheloung and Macbeth, 2002). This has enhanced Australia's reputation over many decades as an exporter of grain of consistently high standard ensured by inspection and certification under the aforesaid legislation that is consistent with the IPPC and importing country expectations. The insects involved are all cosmopolitan species but have the capacity to cause losses through damage to grain especially in major storages where existing pest populations are suppressed under an ongoing programme. Measures to meet export inspection requirements range from insect sanitation in the production and handling chain to rapid disinfestation measures when a grain stream is found to be infested at loading of a ship or container. It is a universal tenet that all grain is considered infested at all times and it is the intensity of the inspection process that determines the allowable tolerance. It has been wide practice for durable commodities to rely heavily on fumigation particularly with methyl bromide. Since the use of this fumigant is being phased out under requirements of the 1987 Montreal Protocol on Substances that Deplete the Ozone Layer (Chapter 10), there is a need to find alternative measures which are as efficacious and cost effective.

For perishable commodities, mainly fruit, vegetables and cut ornamentals, a wide range of quarantine security levels exists in world trade. Pests may be targeted because of their inherent destructiveness or because they are vectors of pathogens, particularly viruses. Countries in which entry of pest-prone commodities are subjected to inspection as a routine practice tend to reject any shipment in which regulated pest species are found and in addition those shipments with live pests that cannot be identified with certainty by the inspector or their immediate support services. The time available for identification of pests in perishable commodities can be as short as a few hours and a policy of rejection as a default action is normal. Small pests such as thrips and mites come within this category. After rejection a shipment may be retrievable by an immediate disinfestation treatment such as bulk fumigation or it may be directed to a processing treatment that eliminates further risk.

Postharvest management practices

Commodities that are judged to have significant risk of carrying pests that cannot be managed as part of the preharvest production system may be acceptable to a market with quarantine constraints if they are subjected postharvest to an approved precautionary disinfestation treatment. This can ensure that the level of quarantine security identified as necessary in a pest risk analysis is achieved. Further product security may be necessary to prevent the occurrence of re-infestation.

The methods available to ensure that any quarantine pest carried on or in produce will be unable to result in establishment in the area of the trade recipient include:

- Cold storage.
- Heating to lethal temperatures.
- Ionizing radiation.
- Modified (controlled) atmosphere storage.
- Fumigation.
- Pesticide applications.
- Miscellaneous other methods.
- Combinations of one or more methods.

Cold storage (Chapter 7) is the simplest of the physical disinfestation methods. It generally relies on refrigeration technology but the use of suitably cold ambient air is an option for phytosanitary pest management of stored grains and similar commodities. The major advantage of cold storage disinfestation is that it is almost invariably a part of normal operational handling and transportation of perishable commodities especially sea transport where transit time is relatively long. This greatly simplifies the export chain. Its disadvantages include the extent of holding facilities for the relatively long treatment times needed to disinfest commodities of quarantine pests unless rapid low temperature freezing is used and cold intolerance in many commodities at temperatures required to eliminate pest risk.

Heating for times lethal to pests (Chapter 8) is another major physical disinfestation method. General practice is to use the highest treatment temperature that avoids heat injury to the commodity thus minimizing the time required of the treatment as excessive time at subthreshold temperatures for heat damage can expose the commodity to quality degradation and ageing processes that are accelerated at higher temperatures. Heating can be done by convection from heated air, conduction from heated water including condensing steam or by radiation such as microwaves. Air and water are the conventional sources currently in use. Heat treatment has become much more feasible as high precision microprocessors and sensors developed towards the end of the 20th century enabled heating of perishable commodities to within narrow limits of insect lethality and avoidance of commodity heat tolerance without unacceptable injury.

Ionizing energy irradiation treatments (Chapter 9) have yet to reach their full potential usage for quarantine disinfestation purposes. A problem arose early when these treatments were incorrectly perceived as a nuclear process based on the use of radioactive cobalt as a source of gamma radiation as the source of ionizing energy. Alternative sources of ionizing energy are X-rays and electron beams that can be generated electrically. These facilities are complex and expensive largely due to the necessity for radiation shielding and added security. Low dose irradiation has the ability to ensure absolute sterility in pests thus providing quarantine security through prevention of multiplication and establishment in the recipient country. A disadvantage is that pests found at inspection will for the most part be alive. The problem can be overcome in several ways. Pre-export pest management and inspections can minimize the likelihood that pests will be found at inspection and treatment certification can be arranged to provide adequate assurance that minimum dose levels were applied. Irradiation is a physical treatment and can provide the same freedom from residues as heat and cold.

Modified atmosphere storage uses atmospheric gases in proportions that are lethal to pests (Chapter 11). Ongoing control of the atmosphere may be required during storage. Atmospheres may be modified by the addition of or even purging with carbon dioxide or nitrogen or depletion of oxygen by combustion of a carbon source or a catalytic process. The main usage of modified atmospheres is for pest control in grain storages and in combination with refrigeration for maintenance of quality in fruits stored between growing seasons in which role pest disinfestation is secondary. The method has potential for wider usage in a disinfestation role but is hampered by the time taken to be effective at normal quarantine security levels. Because the method uses atmospheric gases normally present it avoids chemical residue problems. However, these atmospheres will not support life and suitable safeguards must be in place to protect operational staff.

Fumigation (Chapter 10) is a well-established technology with skilled operators available worldwide. It can be done with relatively simple equipment but where there is an established need permanent facilities can reduce long-term costs. This gives the technology a degree of logistical and economic flexibility matched only by the other chemical methodology, pesticide application. The disadvantages of fumigation mostly stem from its chemical basis. All fumigants are poisons and in recent times environmental, health and public or operator safety concerns have reduced the number of fumigants approved for use and can be expected to continue to do so. Residues in fumigated commodities are less of a problem than for pesticides but can be detected with modern analytical equipment. Physical methods that generate no substances or only those that are already common in the environment have a consequent advantage in markets where consumer preference is for residue-free produce.

The usage of pesticides on edible commodities in a phytosanitary role (Chapter 12) is limited to those pesticides that can be applied without exceeding Maximum Residue Limits (MRL). For fruits and vegetables with internal pests only pesticides with systemic action are likely to be effective. Pesticides generally have a lower risk of phytotoxicity or other injury on fresh horticultural commodities than other disinfestation methods. For grains and other stored products the potential protectants and disinfestants are more numerous but a low MRL and a relatively short half-life are essential. The use of pesticides as postharvest phytosanitary treatments is decreasing as a result of consumer preference for residue-free methods and occasional instances of tainting in susceptible end products such as beer. Advantages for pesticides include low cost application, usually as a dip, inline spray or as a dust additive and possibly more importantly, the residual protection conferred. Some pesticides are chemically inert and act through causing physical damage to pests but they are dusts and can be difficult to apply to some commodities.

Consumer preference for residue-free pest control measures has led to consideration of a number of hitherto unused potential treatment technologies for phytosanitary purposes. Radiofrequency wave energy heating has received some attention (Chapter 8) but other possibilities include light spectrum radiation including ultraviolet, physical impact, negative and positive pressure, atmospheric pressure plasma discharge and ultrasound (Chapter 13). Radiofrequency radiation is electromagnetic energy that generates heat when it interacts with charged particles and polar molecules. Radiofrequency heating has received considerable research attention but has not been used commercially yet. Microbiological use of ultraviolet light is well established as a space and surface disinfectant against pathogens but not larger organisms. Atmospheric pressure plasma discharge has been used industrially for sterilization of equipment against bacteria and may have potential for use against insects (Donohue *et al.*, 2006).

Physical impact is known to be a significant cause of mortality in grain insects (Bailey, 1962) but is not used in other than a contributory role such as 'turning' of grain in vertical storages for other reasons. Physical impact finds use in commodities such as flour which can be disinfested by an Entoleter™ spinning bar in the pathway of a moving stream of the commodity. Hyperbaric chambers that achieve atmospheric pressures in excess of normal have been mooted for disinfestation purposes but are not known to have been used in commercial operations although hypobaric chambers have been used in conjunction with modified atmosphere storage treatments.

Combined treatments may be applied where one alone would cause unacceptable phytotoxic injury. All physical treatments currently in use can cause commodity injury if the dose required to achieve the desired level of efficacy against a quarantine pest is extreme. Where stages of the pest respond differently to a treatment the effects of more than one treatment, each at relatively low levels, in combination may be adequately efficacious. Instances are referred to of combinations of cold and fumigation, heat and modified atmospheres, cold and irradiation and heat and cold.

2 Agricultural Warfare and Bioterrorism Using Invasive Species

One classification of phytosanitary issues and invasive species is by intention. An example at the most innocent level is the frugal business traveller who packs a leftover apple in a suitcase to take home instead of throwing it away. He gives no thought to possible phytosanitary problems associated with that behaviour; on the contrary, he feels he is doing good by not wasting an apple. A greater level of guilt would be the tourist who suspects that it is prohibited to bring live plants into the country (can't recall why) but cannot resist peeling a small orchid off the bark of a tree on the last day of a trip to a tropical paradise and secreting it into the inner pocket of her jacket. The exotic pet store owner on a trip overseas who puts a colourful snail in his pocket, intent on breeding the hermaphrodite for sale so that others may enjoy his unique find commits an arguably greater offence.

Deeper into the depths of delinquency is the marketer who smuggles tonnes of quarantined cut flowers or relabelled fruit from a quarantined source across phytosanitary boundaries to make a quick, but illicit, profit. Entering the realm of true depravity is the rogue owner of a pesticide application operation who accelerates the invasion of the boll weevil across a cotton-growing region in order to increase sales. But the absolute pit of perdition for violators of regulatory provisions to prevent damaging invasive species is for those who use biological agents as weapons to cause harm to the health, economy or collective psyche of a people. These perpetrators can be governments intent on causing harm to another country, groups with political and social agendas, or individuals making a statement.

Agents of Agricultural Targeted Aggression

Although agricultural warfare and bioterrorism have been discussed in the past, they received renewed prominence after the attacks by 14 individuals from Saudi Arabia with five from other countries on targets within the USA on 11 September 2001. The high profile of international terrorism provokes speculation on whether plant pests might be used as a means of aggression and if so, how this might be done. Although plant pest sciences are almost exclusively focused on preventing or managing plant pests and their effects on production and the environment the knowledge base could be used for offensive purposes (Ebbels, 2003). *The Convention on the Prohibition of the Development, Production and Stockpiling of Bacteriological (Biological) and Toxin Weapons and on Their Destruction* encompasses plant pests and pathogens but lacks adequate focus in this regard (CBW, 2005). Possibly more relevant is the informal Australia Group (AG) of more than 34 countries that has developed a Common Control List for export of plant pathogens that could be used in an offensive role. Part of the list nominates bacteria, fungi and genetic elements and genetically modified organisms on which export controls are to be exercised and a further listing of awareness guidelines for these groups (AG, 2005).

Pest-transmitted plant pathogens

Plant pathogens have logistical advantages over arthropod and other plant pests for such purposes and have been actively considered in the past, such as the use of highly pathogenic fungal rusts which predominantly have a windborne means of dispersal (MacKenzie *et al.*, 1985). On the basis that fungi are responsible for 75% of crop plant diseases, Wheelis *et al.* (2002) justifiably regard them as having the greatest potential for use as an act of aggression against crop plants. In this volume we are concerned primarily with management of arthropod pests of plants, including vectors of pathogens, which are of quarantine or otherwise regulated importance and require phytosanitary action. Viral pathogens typically have a history of strain changes that regularly circumvent plant varietal resistance systems and do not respond to chemical prophylaxis. Such pathogens could be expected to respond readily to selection possibly aided by other genetic modification technology to develop strains which are deliverable via vectors, are highly infective, less likely to trigger plant resistance systems and highly pathogenic. Prevention and management of such plant pandemics requires strategies based on vector monitoring and control, resistant varieties and alternative cropping systems and areas.

The introduction of plant-feeding pest insects or mites as an act of aggression has been considered in the past (Garrett, 1996) but is unlikely to be successful in creating significant crop damage in most instances given the complexities of delivery, uncertainty of success and the probable availability of control measures. Arthropod pests require sophisticated mass culture and delivery systems if they are to be used as a destructive force. However, the introduction of a few individuals of regulated quarantine pests, such as the Mediterranean fruit fly, *Ceratitis capitata*, could disrupt trade resulting in significant market and hence, economic, loss in ways analogous to those caused by animal diseases such as foot-and-mouth disease (Wheelis *et al.*, 2002).

A feasible modus operandi would be use of vectors to deliver a pathogen with a high infectivity potential from a relatively small amount of inoculum. Viruses sensu lato are the most appropriate pathogens for intentional vector transmission, especially given recent advances in knowledge of plant disease transmission by vectors. Vectors are mainly arthropods: thrips (Thysanoptera), aphid (Aphididae), whitefly (Aleyrodidae) or leafhopper (Cicadellidae). Other groups including Lepidoptera, Diptera and nematodes are capable of transmitting potentially destructive pathogens. Vectors predominantly have in common a feeding habit involving piercing and sucking. Homopterans, particularly aphids, exhibit the ability to interact with the virus physiologically and in some instances to carry it from one generation to the next (Gray and Banerjee, 1999; Hanboonsong *et al.*, 2002). Staple food crops most at risk include cereals such as wheat, maize and rice, legumes, especially grain legumes, solanaceous cultivars, particularly potatoes and tomatoes, sugarcane and most fruit including citrus, papaya and bananas.

The threat to humans by bioterrorism is often represented by organisms that are already spread worldwide but must be weaponized, requiring a high degree of technical sophistication. Examples are the September–October 2001 anthrax attacks via the postal service on several targets within the USA (Matsumoto, 2003). It is unlikely that these attacks could have been carried out without uncommon knowledge, skills and equipment. The attackers themselves must take precautions to prevent themselves from becoming victims. Even if the attack is to be suicidal the attackers must prevent succumbing to the infectious agent before being able to carry out the attack, which may take place considerable time after obtaining the infectious agents. Biological attacks to agriculture are simpler to carry out than attacks on humans and may be done with rudimentary knowledge and training without exposing the perpetrators to danger by organisms that are usually no threat to humans (Casagrande, 2000).

Cameron *et al.* (2001) mention 21 incidents worldwide since 1952 that could be considered substate acts of terrorism against agriculture. Most were unsophisticated and lacked significant effect and could be considered closer to product tampering for revenge or financial gain than terrorism designed to influence the general public. Some were hoaxes: a threat involving foot-and-mouth disease in Australia in 1984 generated considerable alarm although there was no evidence that the perpetrator had access to the virus. By and large these incidents have been low level with minimal consequences.

Another incident with an insect occurred in 2003 in New Zealand. The Australian painted apple moth, *Teia anartoides*, has been targeted in an eradication programme in Auckland since its discovery there in 1999 (Suckling *et al.*, 2006). A controversial spray programme uses aerial applications of *Bacillus thuringiensis*, while monitoring of the pest includes pheromone traps baited with live apterous females. An anti-spray activist announced intentions to deliberately spread the insect in protest of the sprays and then some females went missing from traps. As a response studies were carried out on sexually sterilizing the females with irradiation before placing them in traps.

Pests as destructive agents

Sovereign states have used and studied attacks via biological agents with many programmes being well documented (Cameron *et al.*, 2001). Like much of warfare, programmes based on biological agents often escalate because of intelligence on what the 'other side' might be up to. In time of war preserving the agricultural economy and environment often take a backseat to hasty decisions deemed at the time to be of utmost importance. The brief foray of the Colorado potato beetle, *Leptinotarsa decemlineata*, into readiness for battle is one of those instances (Fig. 2.1). Garrett (1996) suggests that interest in the beetle as a biological warfare agent against the staple crop, potatoes, during World War II might stem from the finding of 'abnormal features' in at least one instance with Colorado potato beetle collections west of London in 1941. The beetle, native to North America, was found in England as early as 1901 and France and Germany in 1914 (EPPO, 1992). Although there are no verifiable records of the beetle being released in an offensive capacity during that war, France, England, the USA and Germany were all caught up in intrigue with the possibilities. Germany released at least 54,000 beetles south of Frankfurt to study its weaponization capabilities, and the USA shipped at least 15,000 Colorado potato beetles to England in 1942 for militarization studies (Garrett, 1996). In 1944, a 'devastating' infestation of the beetle was reported in Germany (Lockwood, 1987); it is not known if this was a natural infestation, a consequence of the Germans releasing thousands of beetles into their own lands, or possibly even the result of Allied action.

Instead of being used purposely in battle, invasive species may be unintentionally transmitted by migrating armies to the detriment of the enemy, the invading army's country or allies, or neutral countries. An example is the probable introduction of the West Indian drywood termite, *Cryptotermes brevis*, to Australia during the Pacific War (Heather, 1975). Under desperate conditions of warfare scant attention may be paid to invasive species. Modern armies

Fig. 2.1. The Colorado potato beetle, *Leptinotarsa decemlineata,* was studied as a weapon on both sides during World War II, although it is doubtful that the pest was ever used offensively (Source: photograph by Scott Bauer, USDA-ARS).

withdrawing during periods of relative calm have returning equipment cleaned and inspected to prevent transport of invasive species back to the homeland.

Accusations of states deliberately infesting 'enemy' countries with invasive species abound but for the most part are unsubstantiated and can often be demonstrated as false (Lockwood, 1987; Wheelis, 2004). A recent accusation is that the US military brought the western corn rootworm, *Diabrotica virgifera virgifera*, to Serbia in the early 1990s as a weapon in the military conflict there. The insect probably arrived years before the USA became involved in that conflict and most likely accompanied commercial imports by air from infested areas in the USA (Hostetler, 2002).

Malfeasance in this case is hard to believe; the western corn rootworm cannot be considered a potent weapon, and its spread throughout the region to neutral countries as well as US allies involved in the military operation in Serbia would be obvious to anyone considering using it as a weapon. However, given the secrecy, level of deceit and 'dirty operations' that have been discovered in some cases (Shultz, 1999), scepticism about government innocence is understood. Governments of technologically advanced countries have many resources at their disposal, including cover and legitimacy as well as abundant funds, materials and trained researchers to develop new and frightening offensive technologies and the means to deliver them (Wheelis, 2002). Indeed, during the Cold War the USA maintained huge stockpiles of anti-crop biological weapons and the Soviet Union developed the capacity to produce large amounts on short order (Alibek, 1999; Moon, 2006). On the other hand, precipitous action based on hasty and inaccurate assessments of the imminent danger that a government poses via 'weapons of mass destruction' can be very disastrous.

In 1989, there were allegations that some fertile Mediterranean fruit flies were deliberately released in California to protest against the use of aerial spraying of malathion as part of an eradication programme against the insect (Cameron *et al.*, 2001) although no evidence confirming these allegations surfaced. The number of fertile flies captured in traps in 1989 was 235 versus 54 the year before and 44 the year after. Carey (1991) argued that trends in Mediterranean fruit fly finds were symptomatic of establishment and that detection, exclusion and eradication protocols would need to be changed. Changes were made, and Mediterranean fruit fly finds decreased from that period. In 2005, 33 adult Mediterranean fruit flies were found in California using a higher trapping density (over 20,000 total traps now) than before. Spinosad replaced malathion for aerial spraying in California.

Defensive Measures

Efforts to protect agriculture from bioterrorism may enhance traditional regulatory efforts to protect agriculture from invasive species. Increased surveillance, inspection and vigilance directed towards terrorism should also increase the chances that invasive species entering through non-terrorist avenues will also be detected. However, there is concern that in some cases increased attention to terrorism is reducing attention to the entry of invasive species not connected with terrorism. The US Government Accountability Office issued a report concluding that since agricultural inspection duties were moved from the Department of Agriculture to the Department of Homeland Security 'challenges exist that increase the vulnerability of U.S. agriculture to foreign pests and diseases' (USGAO, 2006).

Invasive species cause tremendous losses regardless of how they were introduced. Pimentel *et al.* (2002) estimate that exotic arthropods and plants in the USA cause annual losses of US$20.1 and 34.1 billion, respectively, whereas the destruction and clean-up of all seven buildings of the World Trade Center in New York in 2001 caused a one-time loss of US$27.2 billion, although major additional losses related to this event are expected (Looney, 2002). The guard against invasive species entering through non-terrorist routes should not be lowered because of vigilance against terrorism.

3 Plant Regulatory Organizations

Regulatory plant protection demands cooperation at many levels from local to international to be effective. The interests of stakeholders, such as growers, importers, consumers and environmentalists, often compete. Interested parties may find themselves on both sides of complex issues, such as domestic growers of one type of commodity that also have plantings of this commodity in quarantined countries, and consumers who desire an exotic commodity but will not purchase it unless it satisfies certain criteria, which could be organic, 'fair trade' or from sustainable agriculture. Organizations that regulate phytosanitary issues affect virtually all growers and consumers today. Even subsistence farmers who do not participate in trade may be affected by regulatory organizations, as they can change the agricultural, social, economic and environmental well-being of the hamlet in which they live.

International Organizations

The International Plant Protection Convention (IPPC) is an international treaty in force since 1952 to prevent the introduction and spread of pests of plants and their products and promote appropriate pest management measures. Where deemed appropriate provisions of the IPPC may be extended to articles capable of harbouring or spreading plant pests, such as soil, packaging, conveyances and storage facilities.

Over four-fifths of nations are currently parties to the IPPC, which recognizes that phytosanitary measures should be scientifically justified, transparent and applied so as not to constitute disguised trade barriers. The IPPC provides a framework for the development of international phytosanitary standards and the application of phytosanitary measures. Parties to the IPPC agree to cooperate on information exchange and on the development of standards and to fulfil the objectives of the IPPC in their own countries according to their competencies. They also agree to promote the provision of technical assistance to other contracting countries, especially those representing economically disadvantaged areas, to facilitate the implementation of the IPPC. Provisions for settling disputes concerning phytosanitary quarantines and measures are also provided.

The governing body of the IPPC is the Commission on Phytosanitary Measures (CPM), membership of which includes all contracting parties. The CPM is supported by a secretariat which resides with the Food and Agriculture Organization of the United Nations (FAO). The Secretariat assists the CPM in reviewing the state of plant protection in the world and the needs for phytosanitary actions. The CPM establishes its work programme and adopts International Standards for Phytosanitary Measures (ISPM; Table 3.1). It also establishes non-binding procedures for settling disputes among contracting parties. Subsidiary bodies may be established by the CPM to perform other tasks as may be necessary for the proper fulfilment of the objectives of the IPPC. It also cooperates with other relevant international organizations on matters covered by the IPPC and adopts guidelines concerning the establishment of regional plant protection organizations (RPPOs).

There are currently nine RPPOs that concentrate on phytosanitary, and sometimes animal health issues as well, within their respective regions and coordinate with the IPPC and national plant protection organizations (NPPOs) to gather information, promote harmonization and implement phytosanitary measures (Table 3.2). The agreement for a tenth RPPO representing the Middle East and northern Africa has not yet been ratified by a sufficient number of countries to come into force. Some countries may be members of overlapping RPPOs. For example, Mexico is a member of the Caribbean Plant Protection Commission (CPPC), the Organismo Internacional Regional de Sanidad Agropecuario (OIRSA; comprising Central America and portions of the Caribbean and Mexico) and the North American Plant Protection Organization (NAPPO).

The Convention on Biological Diversity (CBD), an international treaty in force since 1993, expands the management of invasive species beyond the protection of plants and plant systems. It calls on member countries to 'prevent the introduction of, control or eradicate those alien species which threaten ecosystems, habitats or species'. This convention has been very popular among United Nations members, being ratified by all except Andorra, Brunei Darussalam, Iraq, Somalia and the USA.

The World Trade Organization (WTO) Agreement on the Application of Sanitary and Phytosanitary Measures (SPS Agreement) came into force in 1995 for most of the 151 current WTO members. The SPS Agreement provides a uniform interpretation of measures governing food safety and plant and animal health regulations and is applicable to all like measures affecting international trade. These measures include any applied to protect animal or plant health within a member's territory from the entry, establishment or spread of pests (broadly interpreted to include diseases and weeds). Regulations must be based on scientific principles and evidence, and should be applied only to the extent necessary to protect human, animal or plant health. They should not discriminate between countries where similar conditions prevail.

Table 3.1. International Standards for Phytosanitary Measures (ISPM) approved by the International Plant Protection Convention (Source: IPPC, 2007).

Number	Year of current version	Title
ISPM 01	2006	*Phytosanitary Principles for the Protection of Plants and the Application of Phytosanitary Measures in International Trade*
ISPM 02	2007	*Framework for Pest Risk Analysis*
ISPM 03	2005	*Guidelines for the Export, Shipment, Import and Release of Biological Control Agents and Other Beneficial Organisms*
ISPM 04	1995	*Requirements for the Establishment of Pest Free Areas*
ISPM 05	2005, 2007	*Glossary of Phytosanitary Terms*
ISPM 06	1997	*Guidelines for Surveillance*
ISPM 07	1997	*Export Certification System*
ISPM 08	1998	*Determination of Pest Status in an Area*
ISPM 09	1998	*Guidelines for Pest Eradication Programmes*
ISPM 10	1999	*Requirements for the Establishment of Pest Free Places of Production and Pest Free Production Sites*
ISPM 11	2004	*Pest Risk Analysis for Quarantine Pests including Analysis of Environmental Risks and Living Modified Organisms*
ISPM 12	2001	*Guidelines for Phytosanitary Certificates*
ISPM 13	2001	*Guidelines for the Notification of Non-compliance and Emergency Action*
ISPM 14	2002	*The Use of Integrated Measures in a Systems Approach for Pest Risk Management*
ISPM 15	2002	*Guidelines for Regulating Wood Packaging Material in International Trade*
ISPM 16	2002	*Regulated Non-quarantine Pests: Concept and Application*
ISPM 17	2002	*Pest Reporting*
ISPM 18	2003	*Guidelines for the Use of Irradiation as a Phytosanitary Measure*
ISPM 19	2003	*Guidelines on Lists of Regulated Pests*
ISPM 20	2004	*Guidelines for a Phytosanitary Import Regulatory System*
ISPM 21	2004	*Pest Risk Analysis for Regulated Non-quarantine Pests*
ISPM 22	2005	*Requirements for the Establishment of Areas of Low Pest Prevalence*
ISPM 23	2005	*Guidelines for Inspection*
ISPM 24	2005	*Guidelines for the Determination and Recognition of Equivalence of Phytosanitary Measures*
ISPM 25	2006	*Consignments in Transit*
ISPM 26	2006	*Establishment of Pest Free Areas for Fruit Flies (Tephritidae)*
ISPM 27	2006	*Diagnostic Protocols for Regulated Pests*

Member countries are encouraged to use international standards and recommendations established by the bodies that set international standards (the IPPC in the case of plants) and may use measures which result in greater levels of control given scientifically defensible justification. They are encouraged to participate in the development of international standards and to use risk assessment as the basis for their measures where standards do not exist or

Table 3.2. Regional plant protection organizations.

Organization	Geographical and political area
Asia and Pacific Plant Protection Commission (APPPC)	Asia, Australia, New Zealand, Oceania and France for French Polynesia
Caribbean Plant Protection Commission (CPPC)	Caribbean island nations, northern South America, some Central American nations and parent nations of Caribbean island territories
Comité Regional de Sanidad Vegetal para el Cono Sur (COSAVE)	Argentina, Brazil, Chile, Paraguay, Uruguay
Comunidad Andina (CA)	Venezuela, Colombia, Ecuador, Bolivia, Peru
European and Mediterranean Plant Protection Organization (EPPO)	Almost all of Europe and a few countries of northern Africa and Middle East
Inter-African Phytosanitary Council (IAPSC)	All African countries except Morocco
Near East Plant Protection Organization (NEPPO)[a]	Many Arab countries, Iran, Malta, Turkey, Pakistan
North American Plant Protection Organization (NAPPO)	Canada, Mexico, USA
Organismo Internacional Regional de Sanidad Agropecuario (OIRSA)	Central America, Mexico, Dominican Republic
Pacific Plant Protection Organization (PPPO)	Island countries of Oceania, Australia, New Zealand and France and the USA for their island territories

[a] The Near East Plant Protection Organization has not yet been ratified.

measures need to deviate from agreed standards. Article V.3 of the IPPC (2007) states that countries undertake not to require consignments to be accompanied by phytosanitary certificates inconsistent with the model set out in the Annex to the Convention.

In the WTO, a specific food safety or animal or plant health requirement established by one country that leads to a trade restriction can be challenged by another country, if the latter believes that there is not sufficient scientific evidence supporting the need for the restriction. Challenges to phytosanitary trade barriers have been made a number of times since the SPS Agreement came into force. Failure of a country to accept a decision on a WTO dispute may result in retaliatory trade practices by the affected country.

Roberts and Krissoff (2004) reviewed the WTO role with respect to the SPS Agreement and examined the change in the resulting balance of barriers from trade protectionist to valid phytosanitary based. They found that for fresh fruit, 60% of (non-obligatory) notifications to WTO related to phytosanitation compared to those for food safety issues and for vegetables (about 50%). Between 1995 and 2000, 32 complaints were referred to the WTO Committee for dispute resolution by more than 15 countries with the support often of many other countries. These complaints involved 18 respondent countries, an indication of the widespread use made of the facility. Complaints were in various stages of resolution from completed to about to commence. The SPS Agreement had not resulted in an expected flood of protectionist SPS measures supplanting traditional trade barriers so that the Agreement has been partly successful in its intended deterrence of regulatory protectionism. This has resulted in increased export opportunities for some countries. However, it has highlighted the need for relevant standards and some countries have suffered from their lack of technical expertise and facilities to develop responses to phytosanitary barriers required to initiate trade to new areas.

National Organizations

The IPPC obligates each contracting party to identify its official NPPO. Of course, countries that are not parties to the IPPC may also have official plant regulation and protection agencies. Responsibilities of an NPPO include:

- Inspection of regulated plants, plant products and other items moving in international commerce to prevent the introduction and spread of pests.
- Surveillance of cultivated and wild plants and plant products for pest infestations, control of these infestations and transparent reporting of plant pest problems.
- Issuance of phytosanitary certificates (Appendix I) to indicate that consignments of regulated articles meet phytosanitary requirements of the importing party and maintain the integrity of certification until export. Section 2 of ISPM 12 (*Guidelines for Phytosanitary Certificates*) states that phytosanitary certificates should not include reference to pesticide residues or commercial information (IPPC, 2007).
- Conduct pest risk analyses, which are assessments of the danger of introducing pests that an imported item might present to the importing country.
- Disinfestation and disinfection of regulated plants, plant products and other items moving in international commerce to meet phytosanitary requirements.
- Protect areas endangered by quarantine pests using quarantines and other regulatory mechanisms.
- Designate and maintain 'pest free areas', 'areas of low pest prevalence' and other regulated phytosanitary systems.
- Ensure that staff are up to date with respect to phytosanitary knowledge and procedures.

Most countries have agencies tasked with providing plant protection guidance and services within their own borders and among adjacent countries and trading partners. The degree of phytosanitary expertise and regulation varies widely among countries. Some NPPOs may conduct research into phytosanitary problems. Unfortunately much of that research may not be readily available to the scientific community as there may be little impetus for regulatory officials to publish research results, or the research may not be considered appropriate, or of sufficient novelty for some scientific journals, although with the proliferation of scientific journals at all levels the last point is probably not valid.

Changes in the perception of threats a nation faces may lead to shifts in resources and attention away from agricultural threats. For example after the attacks of 11 September 2001 in New York and Washington, DC, more than 1800 agricultural specialists tasked primarily with the detection of quarantine pests and possible invasive species were moved from the Department of Agriculture to the new Department of Homeland Security in an effort to consolidate port inspections for all regulated items, such as illegal drugs and agents of terrorism. The US Government Accountability Office (USGAO, 2006) realized that management and coordination problems in this reorganization increased the vulnerability of US agriculture and has recommended further changes to improve inspection for agricultural pests.

Larger countries may have subnational plant protection organizations. For example, the National Plant Board in the USA is made up of the plant pest regulatory agencies of each of the states and Puerto Rico, and is divided into four regional plant boards. The plant boards and individual state plant protection organizations function much as the NPPO, but usually on a more local level. The US National Plant Board produces special reports on issues relevant to plant protection, such as a 1999 general review of the US NPPO plant protection efforts (NPB, 1999).

Professional Societies

Professional societies are useful for developing relationships among workers in a narrow discipline, by promoting high standards of professionalism, providing a conduit for dissemination of information and scientific publications and a forum for discussion, speaking with a unified, knowledgeable voice on relevant issues, and increasing public awareness of the discipline and its significance. There are many professional societies with 'plant protection' as a major concern, but this usually means general crop protection designed to produce or maintain a harvest, not as a regulatory endeavour related to trade and the prevention of the spread of invasive species. These societies often have plant regulatory professionals as members and may publish regulatory articles in their journals.

The Society for Regulatory Plant Protection started in the USA in 1994. It was heralded as an independent organization to foster professionalism, communication and unity of regulatory plant protection workers. It had as one of its main goals the consolidation of regulatory plant protection knowledge and the definition of acceptable standards and aims. The first issue of its newsletter, *Plant Protection Connection*, issued in March 1994 was an ambitious eight pages, with a mission statement, a list of regular features to be included in future issues, a book review, pest updates and a cartoon (Fig. 3.1). Unfortunately the society became inactive after issuing its fifth newsletter in March 1997.

Four regional chapters of the Horticultural Inspection Society exist in the USA to promote high standards of and disseminate information about plant inspection and develop professionalism among members. Annual meetings are held with presentations on pest problems and emphasis on inspecting for them.

Fig. 3.1. Prophetic cartoon in the penultimate issue of *Plant Protection Connection* (February 1995) (Source: courtesy of Tim Lockley).

4 Managing Risk of Pest Introduction, Establishment and Spread in a Changing World

The nature of phytosanitation is pest risk management. Trade with zero risk of pest transport is not an achievable goal. Outright prohibition of trade will not prevent the transport of invasive species and goes against international conventions supporting liberalization of trade. The phytosanitary risks associated with trade prohibitions are frequently greater than the managed risks of regulated trade because of the likelihood of contraband, which will be phytosanitarily unsupervised (Grode and Christian, 2002).

There are obvious advantages in being able to categorize non-indigenous pests with a high probability of successful establishment. If this can be done in a way that permits a proactive approach it will be more successful than relying on interception or quarantine barriers resulting from specific pest risk analyses alone. The US Committee on the Scientific Basis for Predicting the Invasive Potential of Nonindigenous Plants and Plant Pests has reviewed this approach in detail with respect to plants, pathogens and pests (Anon., 2002). It concluded inter alia that there are no known broad-based scientific principles or procedures applicable and that, currently, there is a lack of readily accessible, comprehensive knowledge relative to the problem.

Pest risk assessment, an evaluation of the probability of the introduction and spread of a pest and of the associated potential economic and ecological consequences, is a primary component of pest risk management, the evaluation and selection of options to reduce that risk (IPPC, 2007). Prediction of the possible establishment and dispersal of pest species is an essential component of pest risk analysis and is a useful tool for producers in the selection of favourable market and commodity relationships (Baker, 2002). Ebbels (2003) discusses the history of dispersal and establishment of invasive species from the earliest recorded times to the modern era of air transport. Worner (2002) and Baker (2002) reviewed prediction of the invasive potential of exotic pests and the limits to potential distribution, covering biology, climate matching, statistical techniques, modelling, geographic information systems and bioclimatic indices. Baker

(2002) segregated the factors into abiotic, biotic and intrinsic categories as a way of reviewing their effects in any risk analysis. Availability of increasing computational power continues to improve the value of predictions.

Influence of World Climate Change

Climate is a dynamic system and always has been in a state of change in terms of geographic time spans. The rate of change is never constant but in recent decades an accelerated increase in temperature, the principal climatic factor, has become apparent. The mean rise throughout the 20th century worldwide has been assessed as 0.7°C, of which 0.5°C occurred during the last three decades (Anon., 2006a). The UN Intergovernmental Panel on Climate Change concluded that most of the warming in recent history is due to anthropogenic increase in 'greenhouse gases' in the atmosphere (McCarthy *et al.*, 2001). There is scientific consensus on this issue (Pielke and Oreskes, 2005) although politics has tended to sow confusion in some countries (Lovell, 2006).

Already there is a high degree of uncertainty concerning future predictions of invasiveness and geographical distribution of pests but this uncertainty is increased by the fact that recent global climate change is in large part due to human activity. Future human activities and possible attempts to curb global warming could further increase uncertainty in these predictions. It might behove regulatory agencies to be more liberal in their predictions of pests spreading from warmer to cooler regions of the world and adjust regulatory actions accordingly.

Compared to the tolerated biological temperature range and variance that could be expected for most pests, 1°C is a relatively small increment. However, a worldwide variation of 1°C can be related to seasonal temperature ranges, rainfall, sunlight and sea levels in ways that greatly enhance the effects (Anon., 2006a).

Another source of temperature change is urbanization, which can modify local microclimates through energy expenditure and cultural management of the environment. A common example of the effect of urbanization on invasive species is the survival and high pest status of tropical cockroaches during cold winters in heated buildings far outside tropical limits. Both global warming and urbanization have implications for prediction of trade-related pest dispersals and establishments discussed further in this chapter, especially in the rate of change data required when modelling risk.

Predictions from Pest Incidence, Hosts, Biology, Ecological Requirements and Behaviour

Pests that are widespread and have multiple hosts within their area of origin are likely to have a degree of ecological adaptability which will enable them to establish elsewhere in regions of broadly comparable climate given the availability of suitable hosts. This adaptability is likely to be a characteristic of the inherent biological and behavioural make-up of a species. A species that is a pest of a host native to its region of origin is a possible risk to related hosts in a new region, especially hosts that fill the same ecological niche.

Published host lists are valuable in predicting areas vulnerable to invasives and are becoming more widely available for major pest species through the Internet. Examples are the Mediterranean fruit fly, *Ceratitis capitata*, with more than 260 hosts worldwide (Thomas *et al.*, 2005), *Bactrocera dorsalis* with 117 Asian hosts recorded by Allwood *et al.* (1999) and a Pacific Tephritidae host list is currently being compiled (A. Allwood, personal communication). For maximum benefit these should not simply be lists but contain relevant information on the scale and extent of infestation and the variety, growth stage and condition of each host cited, where possible. The requirement for validation of records is paramount.

Elith *et al.* (2006) compared 16 modelling methods using 'presence'-only data for 226 species involving plants and vertebrates but not insects. Presence-only data were effective for modelling existing species distributions for many districts and regions with the models varying in performance and novel models having better performance than older models. However, it is necessary to assign a level of reliability to basic records used, as the location of collection given on a specimen label can represent an interception, not a record of establishment. A similar approach was used by Worner and Gevrey (2006): a self-organizing map (SOM) system to determine pest species assemblages for global regions based on a presence or absence for 844 species. Individual pest species were then ranked in terms of risk of invasion based on the assemblage that was characteristic of each geographic region. These approaches are valuable general indicators of likelihood of establishment of a new pest in an area but should not be used as the sole basis for decision. The Port Information Network of the US Department of Agriculture (USDA) Animal and Plant Health Inspection Service (APHIS) has a large potential data set for use in this way (Anon., 2002) together with corresponding quarantine service databases of other countries.

Mathematical Modelling

Modelling represents the best approach in utilizing the immense volume of information ancillary to the potential for pest introduction and establishment. Climate, predominantly temperature, pest biology and hosts are the key parameters of success of a pest. Volume of trade in a host commodity has an influence on risk of introduction as was studied for sweetpotato weevil, *Cylas formicarius elegantulus*, for the southern USA (Addo-Bediako *et al.*, 2007).

Linear temperature models

These are possibly the simplest models and are estimated typically as degree-day requirements for development of an individual of a species (Worner, 1994; Baker, 2002). The relationship is usually linear about the optimum for the species but becomes nonlinear as higher and lower thresholds are approached. If a pest species can be reared in the laboratory its temperature requirements can be estimated experimentally and may be available already in published literature. Care must be exercised to ensure that the experimental population is representative of the intraspecific variation. Temperature alone, while being a major driver of development especially in poikilotherms, is complex in its effects and interacts with other factors. Degree-day requirements alone are of limited use in determining the suitability of a new locality for an invasive species. A major problem is that temperature records are frequently available only as daily maxima and minima and the site of temperature measurement may not be representative of the ecological niche that the species might occupy. However, weather station records are likely to be the only data available and in a suitably robust model can provide reasonable certainty that a pest cannot persist in an environment under investigation. In Chapters 7 and 8 we discuss some of the physiological effects of extreme high and low temperatures on pest species in relation to phytosanitary treatments. This information is relative to geographical distribution as well.

Capacity for increase

Capacity for increase of a species is a worthwhile measure of the likelihood that a pest will prosper in a new environment and has achieved prominence as valuable input into more complex models. Intrinsic parameters influencing success of an organism in a new location include requirements of temperature, humidity, food supply and phenology. Of these, temperature is fundamental as it impacts directly on pest survival and multiplication and through interaction with other parameters. Potential success can be assessed in terms of fecundity and longevity under given conditions that together govern the capacity for increase of a population.

A widely used integrated index for estimating ecological success of a species is the intrinsic rate of natural increase of a population, r_m, also known as the Malthusian parameter, the innate capacity for increase or various combinations of these (Southwood and Henderson, 2000). It refers to a population of stable age distribution in an unlimited environment and hence represents the maximum potential of the species against which the limitations of any environment can be compared. Under defined conditions the value for r_m may be estimated by means of a life table study from the approximation:

$$\sum e^{-r_m x} l_x m_x \rightarrow 1$$

where x is the pivotal age (mean age of females in the cohort at the birth of female offspring), l_x the proportion of females alive at the commencement of each age interval, and m_x the number of female offspring produced per surviving female within each age interval. This model assumes a roughly equal ratio of females to males. Another approximation, the capacity for increase, r_c, proposed by Laughlin (1965) is widely used by insect ecologists.

Simulated laboratory experiments

Relevant laboratory experiments are usually related to host susceptibility or determination of pest survival temperature and humidity thresholds representative of the new environment. Susceptibility of potential hosts from the new area can be tested in the laboratory but results are not always indicative of experience in the field. A standard for the determination of non-host status to tephritid fruit flies has been adopted by the Asia and Pacific Plant Protection Commission (APPPC, 2005). Laboratory tests on hosts can encompass choice/no-choice situations and test the ability of a potential host to support development of the pest through to a viable ensuing generation. Results can show the possibility that a host will support establishment alone or in association with other seasonally occurring hosts, but the presence of the hosts alone cannot be taken as evidence that the pest would establish in a new environment. Upper and lower climatic thresholds would be better established using an integrated biological index such as *r*.

Holistic GIS models

The US Geological Survey (USGS, 2000) defines a geographic information system (GIS) as 'a computer system capable of capturing, storing, analyzing, and displaying geographically referenced information; that is, data identified according to location' and considers that 'Practitioners also define a GIS as including the procedures, operating personnel, and spatial data that go into the system'. A GIS system can integrate data from a range of sources leading to a conclusion about the relationship.

CLIMEX is essentially a GIS modelling system and has predominant relevance to the problem of predicting the movement and establishment of invasive pests (Sutherst and Maywald, 1985). Such models extrapolate existing trends to produce models of possible future scenarios highly relevant to problems of invasive pests in a changing world. It should be realized that change has always been a characteristic of the world environment and that future directions can be predicted but not assuredly known from past history.

The CLIMEX model has been used to predict potential relative abundance and distribution around the world for many pests including the animal parasites *Boophilus*, *Rhipicephalus* and *Amblyomma* ticks and *Haematobia* spp. flies. Also it has been used to model the potential worldwide distribution of Colorado potato beetle, *Leptinotarsa decemlineata* (Baker *et al.*, 2000), Queensland fruit fly, *Bactrocera tryoni* (Yonow *et al.*, 2004) and Mediterranean fruit fly, *C. capitata* (Vera *et al.*, 2002). The fruit fly models benefit greatly from laboratory studies on development and host records but field population inputs are based mainly on trapping studies and surveys using male-only lures. This opens them to questions involving provenance of trapping studies with respect to optima for a species and behaviour in species such as *B. tryoni* which overwinters as adults and females do not become gravid until rainfall-related conditions in spring occur (May, 1963).

Principal component analyses

This flexible multivariate statistical technique is a way of simplifying a complex set of data by reducing it to a few characteristics. Worner (1994) discusses its use, comparing it favourably with climatographs that are limited to two or at the most three variables, to simultaneously explain the presence or absence of a species. Findings are likely to be indicative trends rather than specific determinations.

Pest Risk Assessment

While there has been international acceptance of the need for harmonization of pest risk assessment as a precursor of quarantine or phytosanitary action for some decades there remain a number of differing models in use. There is reasonable agreement among these models but opportunities exist for further harmonization. Of the several models currently in use internationally some are derived from regional plant protection organizations (EPPO, 2006) or the International Plant Protection Convention (IPPC, 2007). Others were developed by quarantine authorities in countries with established phytosanitary import procedures (APHIS, 2000a; AQIS, 2003). The European and Mediterranean Plant Protection Organization (EPPO) model (Figs 4.1 and 4.2) is typical of the process for undertaking a pest risk analysis (PRA) and management used by many countries throughout the world. It consists of a sequence of questions with yes/no answers followed by subjective assessments of subsequent questions (Appendix II). Holt *et al.* (2006) proposed a probability assessment derived from the scoring system as a quantitative risk assessment system, as a more mathematically rigorous process. However, it is inevitable that many aspects of PRA will be based in large measure on subjective judgements of the most skilled professional regulatory staff available (Devorshak and Griffin, 2002).

Much of this international activity was in response to obligations which are derived from the SPS Agreement (WTO, 2007a) which, in Article 5 requires in detail, the 'Assessment of risk and determination of the appropriate level of sanitary or phytosanitary measures'. Two further key international agreements are the abovementioned International Plant Protection Convention (IPPC, 2007) and the Convention on Biological Diversity (see Chapter 3).

Risk assessment is a widely used practice in agricultural plant protection (Sequeira, 2002; Ebbels, 2003). Rohwer and Williamson (1983) detailed the practice from a USDA Plant Protection and Quarantine viewpoint and attributed modern formalization of the concept to Kahn (1979). A PRA is a structured investigation done to determine the possibility of establishment of unwanted pests in the proposed recipient country of a traded commodity. It is carried out by the recipient country but relies heavily on information from the producer country. It takes into account possible biological, environmental, economic and social impacts of pest entry and establishment and tends to assume that risk is high in the absence of technical evidence to the contrary.

A PRA can range from a simple expert opinion in response to an urgent situation to a more usual formalized multiple-step process such as ISPM 02

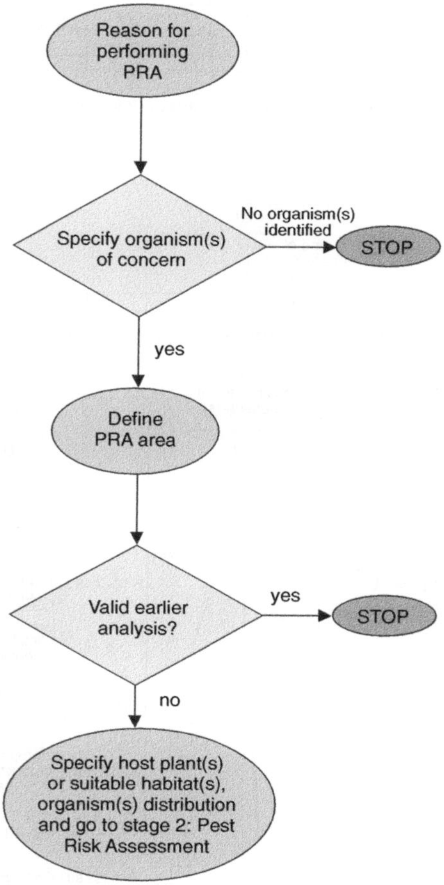

Fig. 4.1. The EPPO model for initiation of a pest risk analysis (PRA) – see Appendix II for a template for a structured assessment (Source: EPPO, 2006; reproduced courtesy of EPPO).

(IPPC, 2007) conforming to requirements of the World Trade Organization (WTO) SPS Agreement. Its purpose is to identify important factors for the formulation of a pest risk management plan that might involve absolute prohibition, risk managed entry or no necessary action. Focus of the analysis may be primarily through a commodity or other pathway of transit in which instance there may be multiple pests requiring consideration or it may be focused on a single pest or pest complex, for example when a new pest is found in a previously analysed pathway. An analysis may be binding at the time it is completed but it must always remain open to modification in response to changes in the knowledge base about the target pest(s), interaction with host commodities and possible effects on faunal and floral biodiversity in the recipient region. Pest risk can be divided into risk of entry and risk of establishment. Risk of entry may be attributable to biological factors including host susceptibility or it may relate to production and handling procedures including transport time. Risk of

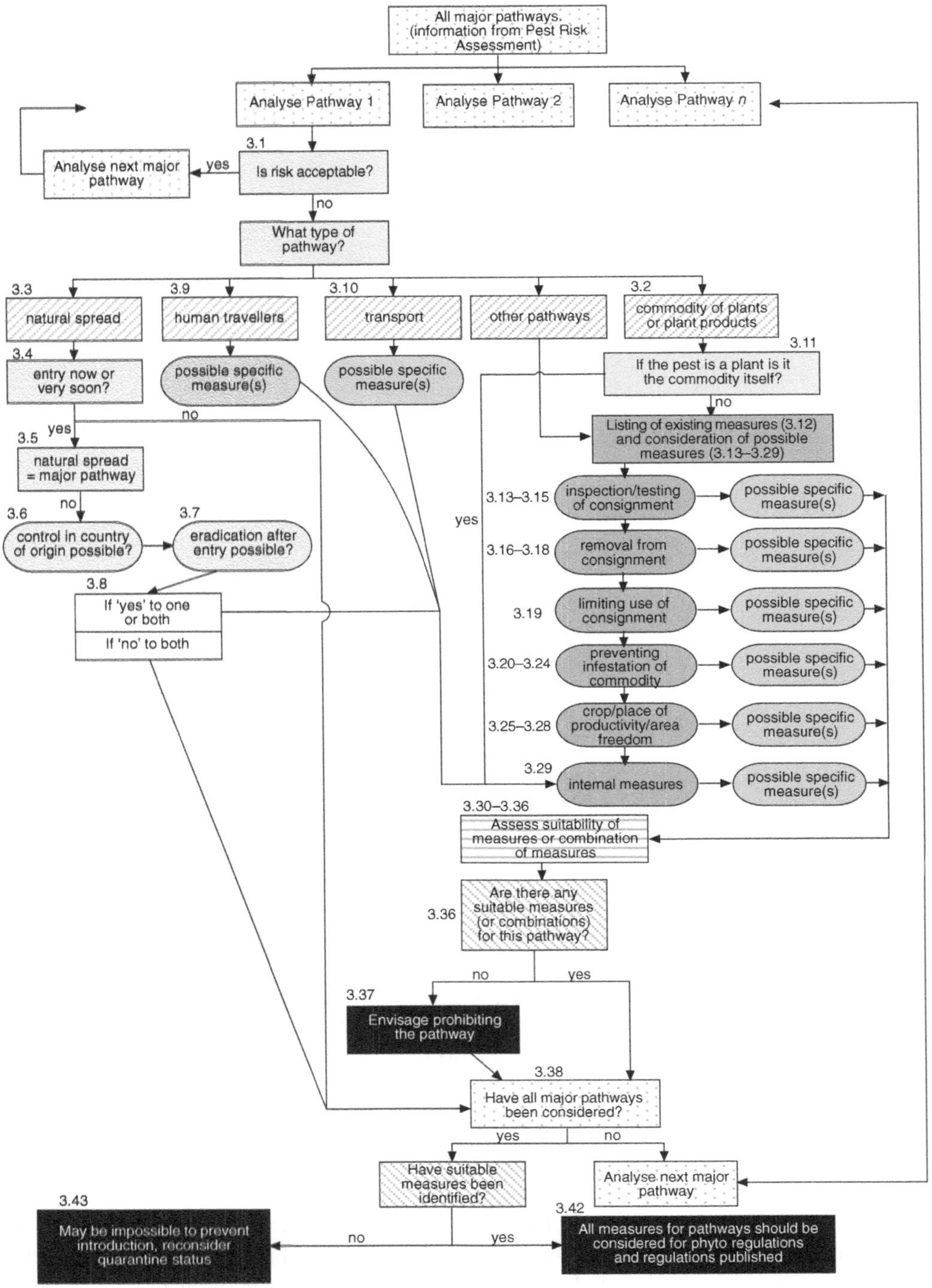

Fig. 4.2. The EPPO model for pest risk management (Source: EPPO, 2006; reproduced courtesy of EPPO).

establishment relates to the interaction of the pest with ecological conditions encountered and covers a broad range of factors from the infestation threshold for success to opportunities to develop and reproduce in the new area.

Factors Relevant to Pest Entry

A number of factors can affect the risk of pest entry. Foremost among these is the relationship of the pest to any of its hosts that may be imported as well as non-host routes of transport.

Pest status

As one of the first actions in a PRA the status of the pest under consideration as a quarantine risk needs to be established both globally and locally within the production region of the trading partner proposing to export the host commodity. It may be possible to do this almost entirely from the published literature on the species or field surveys may be necessary. The IPPC standard ISPM 08 covers means of determination of pest status in an area (IPPC, 2007). Pests of international status will be important both for the harm they might cause to hosts following establishment in a recipient country and for the possible loss of pest free status in exports of that country. Cosmopolitan pests can be awarded quarantine pest status under some circumstances such as where new entries could be disease vectors or where the new entries are resistant to a pesticide whereas indigenous populations are susceptible. Care needs to be exercised with pests apparently of negligible or minor importance in areas to which they are indigenous, as there are many examples in the history of agriculture where such pests can become very serious in a new area in the absence of natural controls exercised in their native environment. Status of a pest with respect to proposed trade is usually a subjective judgement, supported by evidence of its pest potential or the absence of evidence that it will not become a pest in the new environment.

Host–pest phenology interaction

For a pest-focused risk analysis the interaction of phenologies of host(s) and pest is important. Where the commodity for export can be grown at a time of the year when the pest species is inactive, quarantine risk can be much reduced or even reliably avoided. If there is a broad reliable knowledge base on the biology of the pest in the exporting country a decision based on this aspect may be possible. However, in the more usual circumstance where there is a degree of uncertainty, a precautionary approach is prudent, although this may be revised if subsequent scientific research gives reason.

Host status

Accurate documentation and validation of host susceptibility is essential. It is not adequate to undertake a literature search for recorded associations of pests with the commodity to be exported. Similarly, label data on pest specimens in reference collections indicating only its site of collection may not differentiate between incidental association of the specimen with a commodity and a true host–pest relationship. Therefore, published host lists need to be validated at an early stage of market access planning. In the absence of valid host susceptibility studies an import control authority could be expected to assume host susceptibility wherever a reasonable doubt exists so prior studies should be undertaken wherever possible. The Asia and Pacific Plant Protection Commission (APPPC) has published a standard for host testing procedure, in this instance specifically for non-host status against tephritid fruit flies (APPPC, 2005) developed from the New Zealand Ministry of Agriculture and Forestry Standard 155.02.02 (Anon., 2001a). This standard lists host testing stages involving laboratory cage trials with punctured fruit to simulate the best possible chance of successful infestation (stage one) followed by trials with non-punctured fruit if stage one was successful and field cage trials using non-punctured fruit on the host plant, if the second laboratory stage was successful. Emergence of adult flies from the third stage test categorizes the commodity as a potential host. This standard is an advance on the New Zealand standard but the procedure remains cumbersome and inflexible.

Laboratory testing can give misleading results even when the testing is carried through to emergence of viable adults of the next generation, as environmental factors may be involved in host selection in the field (May, 1954). Rearing of a pest from field-collected hosts is generally the most reliable source of host susceptibility but collections need to be made during periods of activity of the pest, preferably over more than 1 year. Also, where ripeness of fruit is involved, maturity must be assessed correctly and the relationship of colour and ripeness determined.

Some fruits with a climacteric pattern of ripening are not susceptible to infestation by pests such as quarantine categorized tephritid fruit flies in the pre-climacteric phase. Examples are bananas and mangoes. Hard green (pre-climacteric) bananas were accepted by a number of countries including Japan, USA and Australia as non-hosts of these tephritid fruit flies and could be admitted or traded internally without treatment requirements against this pest (Anon., 1989; Armstrong, 1994, 2001; Gold *et al.*, 2002). Descriptions of maturity levels in relation to host susceptibility are liable to be confused by commercial ripeness gradings and this is especially so for bananas. Localized precocious ripening at an injury site or as a result of seasonal stress is highlighted by Armstrong (1994) as a problem in defining fruit fly susceptibility in bananas and other fruits such as avocados.

For the purposes of banana susceptibility the term ‘mature green’ used by Armstrong (2001) would equate to ‘immature’ used by the Japanese quarantine service and ‘hard green’ by some others. It is probable that the climacteric point is critical. Better definition of maturity as it applies to pest susceptibility is required together with international harmonization. Examples of other fruits that can be non-susceptible at stages of maturity before ripening include papaya, avocado and mangoesteen. Varietal non-susceptibility to tephritid fruit flies has been accepted for smooth cayenne pineapples (Armstrong, 1994) enabling trade without phytosanitary constraints against this pest group.

Early maturity level was used as one of the components of a multi-stage phytosanitary programme culminating in fruit selection combined with hot water dipping for papaya to be imported to mainland USA from Hawaii (Couey and Hayes, 1986). It was based on the studies of Seo and Tang (1982) and Seo *et al.* (1983), which related benzyl isothiocyanate levels in pre-coloured papaya fruit to tephritid fruit fly susceptibility. This illustrates the importance of consideration of host status as part of the prediction of geographic distribution and establishment of pests already discussed.

A valid host record is one that meets the following minimal requirements:

- Adult pests shall have been reared from infested field-collected host material, or reliably observed in the act of causing commercially or ecologically important damage.
- Both the pest and the commodity species shall have been identified formally in adequate detail by recognized specialists for the groups.
- Factors such as commodity maturity critical to the host requirements of a pest must be accurately defined such as where colour alone does not differentiate pre- and post-climacteric host condition.
- The location and season of collection must be substantiated and in adequate detail to permit validation.
- The host–pest record must have been published in an internationally recognized technical publication or be supported by a statement of required detail from a recognized scientific authority.

The host status of a commodity should be determined by testing, surveys and monitoring of the following parameters:

- Seasonal incidence of the pest relative to time of production of the commodity.
- Host varietal effects on susceptibility to the pest.
- Production and operational procedures which may influence likelihood of infestation.
- Physiological maturity required for commercial trade at the time of harvest.
- Ripeness and maturity especially for climacteric fruits.
- Historical trapping and commodity sampling records of pest populations in the production environment.
- Historical records of any pre-export commodity inspections.

Where a commodity is a favoured host of a pest its susceptibility will usually be well known and accepted as fact. For low susceptibility hosts, the likelihood of infestation needs to be estimated. This can be done by sampling over one or more seasons provided that adequate areas are sampled with respect to diversity of sites and times. Laboratory susceptibility (pest resistance) testing or caged pest trials on the host in the field are alternatives. Care must be taken with caged laboratory or field trials as the pest cannot be given the diversity of alternative hosts which exist in the field (i.e. it is a no-choice or limited-choice test). Host preference assessment can be unreliable where confinement and, often, an unusually high density of the pest induce infestation. An example of this is the recording of French bean as a host of *C. capitata*. Failure to infest under caged conditions is strong evidence of non-host status but could be misleading where prior injury of a particular type is a predisposing factor in the field or where ambient conditions such as light can have an influence on pest activity. The pest individuals used in tests must be at least as fit as field specimens and must be at a suitable stage of development, for example where oviposition is involved females must be gravid and egg viability high. Development of the pest to the next generation is a reasonable test of host susceptibility and the commodity must be held under conditions ideal for the survival of juveniles until adults develop. Testing of the ability of the F_1 progeny to proceed to a further generation is desirable.

Follett and Hennessey (2007) apply procedures for calculating confidence levels (CLs) and sample size requirements for phytosanitary treatments (Couey and Chew, 1986) to the non-host concept and argue that non-host status 'can stand alone as a measure if there is high efficacy and statistical confidence'. Examples from the literature indicate that there is more to confidence in non-host status than statistics. Follett and Hennessey (2007) cite a study where non-host status of 'Sharwil' avocado to Oriental fruit fly, *B. dorsalis*, was demonstrated to the 99.9968% level with a CL of 97.8% (Armstrong, 1991), a very high statistical level of security for any phytosanitary measure. Nevertheless, within a relatively short period after that protocol was accepted avocados satisfying the protocol were found infested with Oriental fruit fly (Liquido *et al.*, 1995). Another study shows an extremely low statistical level of security for non-host status of *Monstera deliciosa* fruit for Caribbean fruit fly, *Anastrepha suspensa* (Gould and Hallman, 2001). This non-host status has not failed in the intervening years. These examples show that statistical measure of non-host status as well as phytosanitary treatment efficacy (Landolt *et al.*, 1984; Mangan *et al.*, 1997) fall short of giving the whole story on efficacy of phytosanitary measures. There is much to consider regarding 'good laboratory practices' when investigating the efficacy of phytosanitary measures (Chapter 6). It can be argued that when phytosanitary measures have failed it has been independent of level of confidence claimed and more related to identifiable incongruities in the research and/or interpretation of it. For example, hot water immersion of mangoes failed even though statistical CLs in the treatment were much higher than 'probit 9' at the 95% CL. Sharp (1988) reports that a total of > 413,000 Mexican fruit fly larvae, *Anastrepha ludens*, and > 365,900 West Indian fruit fly larvae, *Anastrepha obliqua*, were tested with no survivors, the most for any species tested, yet these are the very species for which the treatment seemingly failed in 2000 (Scruton, 2000).

Achieving a high level of confidence (such as 'probit 9') in non-host studies could prove logistically impracticable for many tentative host–pest relationships given that it will be hosts of very low inherent susceptibility which will be the subject of such tests with consequent very low numbers of survivors if any. There will be difficulties in meeting statistical host sampling criteria and could lead to unduly onerous requirements on under-resourced countries wishing to initiate a new export industry. There is also a question as to whether the demands of producing very large numbers of organisms tested can lead to carelessness in the research methodology and/or interpretation. For some prospective non-hosts a combination of a history of not being reported as a host where the commodity and pest exist together or dubious reports of host status coupled with a modicum of successful laboratory and field trails may offer reliable assurance of non-host status. In the rush to harmonization it would not be good to make one-size-fits-all demands on phytosanitary measures.

That said, it is very useful to apply statistical methods to non-host status keeping in mind that statistics cannot be used as a crutch to replace 'good laboratory practices' and that certain levels of statistical confidence should not be standardized as has been the case with 'probit 9' for so many years.

Alternative hosts

Few pests are confined to a single host. Knowledge of other hosts is necessary including those that are not commercial crop plant species. These include species of environmental importance. The analysis must therefore take into consideration other plant species which might be infested should establishment occur in the recipient country. In the assessment of potential hosts in the new area, plants of the same genus or family need to be included in those given consideration. Where the knowledge base on hosts is limited a prudent approach must be adopted towards possible hosts but requirement should not extend beyond reasonable scientific justification.

Pest Incidence in the Area of Origin of the Commodity

Pest importance is normally proportional to its incidence in a production area. This can be expected to govern the likelihood of an infested commodity in trade. Pest incidence can vary from year to year in response to weather patterns in a production area. Where the production area is pest free but adjacent areas can have pest presence, measures must be in place to monitor pest dispersal into the production area and, if necessary, to actively prevent such dispersal. Factors of importance include continuous monitoring with traps throughout the year if appropriate or host sampling programmes during production seasons. For trapping, the density and trap efficacy must be assessed and found adequate, as must the sample size and frequency of sampling programmes.

The apparent absence of a pest of quarantine importance in a production area can be the result of insufficient searching. Therefore absence of records of a pest must be supported by evidence of surveys using valid techniques, adequate in terms of time and scope. Acceptance of absence of a pest is a decision based on risk assessment. Claimed absence of a pest from an export production area can give rise to conflict of opinion and may require dispute resolution by an outside body such as the WTO to ensure that countries with limited scientific resources are not unfairly prevented from participation in international trade.

Pest Free Production Areas and Maintenance of Pest Free Status

Requirements for this status can be:

- Continued pest absence assured because of isolation resulting from ocean, mountain or desert barrier or climatic unsuitability.
- Pest eradicated from production area and maintained free.
- Pest populations subject to active suppression such as by ongoing sterile insect technique (SIT).
- Area under regulatory control.
- Pest incidence monitored by an accepted regulatory authority.
- Absence of prophylactic control measures against the pest.

The reliability of possible pest free production areas is of key importance in a pest risk assessment. Establishment and maintenance of pest free areas is the subject of ISPM 04 and 10 (IPPC, 2007) as discussed in Chapter 5 and elsewhere in this volume. Apart from countries with total pest free status for specific pests, in large countries such as USA, Canada, some South American countries and Australia, naturally occurring pest free production areas will be available for many commodities. These are likely to be geographically isolated and reliably maintainable by quarantine regulation and surveillance. Additionally, pest free areas can be created by eradication programmes followed by regulated monitoring and with pre-planned thresholds and reaction measures should any incursion be detected. Monitoring by trapping or sampling can be established at levels appropriate to the type of pest free area but precautionary pest control programmes are not permissible. However, some flow-on effects will inevitably occur from control programmes against non-regulated pests. Problems may exist with smaller country entities where there are adjacent areas that come under another country's regulatory authority in which the pest incidence is not reliably known or controlled.

Pest Management in the Production Area

Pest management during production of a traded commodity will have a major effect on infestation levels and hence on quarantine risk. Ideally, pest management in non-regulated production areas should ensure an effectively pest free commodity for high-risk pests so that any mandatory postharvest quarantine disinfestation action functions as an insurance measure. Pest management during production can be acknowledged through bilateral agreements between trading partners such as that detailed by Australia and New Zealand for tomatoes (Anon., 1993). Where pest management during production and handling of a commodity can be reliably ensured as in the 'systems approach' (Chapter 5) it becomes a useful element to be considered in any PRA.

Knowledge Base of Pest and Hosts in Area of Origin

PRA requires a knowledge base from which to function effectively (e.g. EPPO, 2007a). Without a comprehensive knowledge base an assessment will always be cautious and precautionary. Where there is uncertainty or absence of reliable technical data, studies may be requested of the exporting trading partner. Some countries, for example Japan, require a detailed submission for certain critical pests before proceeding with their PRA (Anon., 1989). A problem in the use of the knowledge base is again negative proof. When a pest is found or a host is established it is possible to proceed with estimation of the risk. However, when it is not found the extent of the search must be assessed in terms of other knowledge which impacts on the likelihood of presence or otherwise of the pest. This can lead to extensive delays in a decision of approval or continued prohibition of trade where excessive detail is required.

Postharvest Factors

Where temperatures unfavourable to pest development are part of normal operational procedures involving cooling, heating or processing of a commodity which impact on pest survival, the risk of entry of the pest in the commodity will be lessened. If this can be regulated and monitored the effect can be factored into the pest risk assessment with confidence. It may even be possible to use these handling operations as an assurance of adequate quarantine security but it is more likely that they will form significant components of a pest management system as part of a 'systems approach'. Other handling procedures that can contribute to a reduction of pest incidence include washing and coatings.

Culling of infested product at harvest and during packing can further reduce the incidence of a pest in the traded commodity as discussed in Chapter 5. Protection of the product against infestation during packing and transport to the export market, including pest-proof packaging, pest screening of the packing and handling facilities and general security are discussed below. Sampling and inspection may be for quality or phytosanitary purposes and are essential considerations in the assessment of risk of entry of a pest (Chapter 5).

Efficacy of available disinfestation treatments

Most modern disinfestation treatments have been assessed for efficacy as part of their technical development. The efficacy should be known in terms of its mortality effect at a defined level of statistical probability, for example 99.99% at a CL of 95% or probability (P) ≤ 0.05. The required efficacy of a treatment may be determined by national policy, for example for USA 99.9968% at CL 95% for fruit flies, or it may be determined on a case-by-case basis. Long-established treatments such as methyl bromide fumigation could have been approved in the past for use on the basis of extrapolation of efficacy experience from other products without further experimental evidence. Such treatments are at risk from being more severe than necessary. In preparing a PRA, information on the efficacies of available disinfestation treatments can be expected to be available. The efficacy required should if possible not be factored into the PRA until the likely pest incidence in consignments of a commodity is known. Historically, some countries have required demonstration of disinfestation efficacy by the exporter even where it is in use elsewhere in the world or where it is already in use for exports to another country. Acceptance of the principles of harmonization and equivalence in conformity with Articles 3 and 4 of the SPS Agreement (WTO, 2007a) should be followed in the absence of any overriding factor arising out of Article 2.

Transportation and handling factors including security, type, distance and time

The reliability of quarantine security during handling and transport of the commodity is a key component in the risk of arrival of a pest in trade. Pest free shipping can be ensured by inspections prior to loading (Pheloung and Macbeth, 2002). For most field pests, the risk of infestation during postharvest handling and carriage will be low. Pest proofing of packing sheds is feasible and can be made a requirement but not of grain bulks except where hermetic sealed storage is used. Requirement for cool storage at temperatures below thresholds for pest activity will provide good levels of protection against pest infestation during postharvest handling. Packaging can be made secure by required stratagems such as the use of laminated insect-proof package materials or screening of forced-air cooling vents. Where pesticides are required for a disinfestation treatment subsequent residual protection can be expected, depending on the pesticide required. For species such as termites, ants, wasps and bees, risks have increased with the development of containerized shipping in which a colony can be carried accidentally. However, for those social species which require a colony structure the chances of arrival in a reproductively capable state are much reduced, although the bees themselves may carry pests of commercially colonized and feral bees, both of which are depended upon for pollination of agricultural crops.

Other Possible Means of Introduction

Apart from commercial trade a pest may be dispersed with travellers. For example, Liebhold *et al.* (2006) found a continual flow of Mediterranean fruit fly immatures in fruit through southern California airports and concluded that increased baggage screening would help reduce occurrence of infestations of this and other invasive species. Dispersal with travellers is very difficult to monitor, and it is not a priority in most countries, although New Zealand and Australia have aggressive baggage inspection programmes for invasive species (Fig. 4.3). Intentional carriage of a prohibited pest host may be discouraged by the threat of prosecution and effective monitoring such as X-ray of accompanying baggage and the use of animals trained in detection of relevant odours (Fig. 4.4). Given the number of travellers passing into at-risk zones the possibility of operational

Fig. 4.3. Large informative signs in airports are part of an educational facet of an aggressive, integrated effort in Australia to keep invasive species out.

Fig. 4.4. Detection of items of quarantine interest, especially fresh fruit in hand and checked luggage at airports, can be facilitated by use of trained dogs. Australian Quarantine uses beagles, which are highly regarded by travellers and inspection staff alike (Source: photograph courtesy of Australian Quarantine and Inspection Service – copyright Commonwealth of Australia).

failure is always present even if the detection methodology is highly efficient. If a prohibited pest host has a special emotional or ethnic value there will be active attempts to elude detection. Therefore if quarantine regulated trade can make a desired product available in a destination country there will be less likelihood of attempts to elude monitoring for product carried by travellers.

Pests may be distributed naturally (i.e. without human intervention). Natural dispersal of pests can occur by way of wind systems and ocean currents. Migrating birds are a major means of natural dispersal of pest plants. Many pests which can disperse naturally will have already done so. Exceptions will occur where a possible host becomes newly available; or where new strains of a pathogen such as cereal rust have recently evolved; or as climates change and create possible new opportunities for invasive species. These factors need to be recognized as limitations in the assessment of any regulated system intended to identify and minimize risk.

Factors Relevant to Establishment

A number of circumstances can favour establishment of quarantine pests, including availability of hosts and suitability of environments. Also, early detection ability and susceptibility to eradication measures affect the possibility of preventing permanent establishment.

Availability of hosts and suitable environments

For a pest to establish in a new location, appropriate hosts or environmental niches must be available. The available potential hosts may not be the commodity traded if the pest is adapted to a range of possible hosts. However, the hosts must be available at the time of entry or at least within the survival span from entry of the pest. Annual host plants need to be in a susceptible phase of development while perennial hosts need to be at a specific susceptible stage if this is necessary (e.g. fruiting or foliated). The hosts must be within the natural or assisted dispersal range of the pest from the locations of arrival at the relevant time. The necessary coincidence of these requirements means that there are likely to be many more instances of unsuccessful entry and/or establishment of the pest, than there are successful instances. For pests such as ants that do not need a host, availability of a suitable environmental niche is the main requirement. However, here the niche must be unoccupied by a competitor or the pest species must be able to out-compete the existing occupant. As the numbers at introduction will normally be low and must be of the reproductive caste, the chance of successful establishment will be higher if a niche is unoccupied.

Volume of commodity traded

Together with the possible levels of infestation, the volume of a high-risk commodity traded will strongly influence the probability of establishment. Any pest will have density thresholds for successful infestation and establishment. Estimation of possible thresholds has been studied for some pests, especially fruit flies (Landolt *et al.*, 1984; Baker *et al.*, 1990) but for most pests it would be unknown. The history of unsuccessful pest dispersal before regulated quarantine indicates that for many pests the threshold for successful establishment in terms of numbers and frequency can be quite high.

Geographic and climatic suitability of importing region

Disjunct dispersal of pests on world trade routes is well illustrated by the current distribution of cosmopolitan pests such as grain beetles and moths. The Mediterranean fruit fly, *C. capitata*, has a distribution pattern which can be related to centuries-old traditional trade routes but its successful establishment must be related to the need for a particular climatic pattern in many instances, for example, Central America and parts of South America, the Atlantic island of St Helena, parts of South Africa, western Australia but not eastern Australia or New Zealand, the Hawaiian Islands and Florida and California in the USA before eradication. The grain weevils, *Sitophilus* spp., are restricted in distribution so that *Sitophilus zeamais* persists only in tropical regions, *Sitophilus oryzae* in warm temperate regions and *Sitophilus granarius* in cool temperate regions. To classify any one of these weevils as a quarantine pest in an unsuitable region would be an improper estimation of risk.

Efficacy of detection and survey methods

Surveillance of at-risk areas following commencement of trade in a commodity that is host to a quarantine pest is a usual requirement of managed risk. Pest-specific lures and traps must be available or need to be developed. Knowledge is required of the necessary density of trapping sites and the effective life of the lure in each trap. Surveillance staff who are trained in the recognition of the pest are needed. All of these requirements comprise an ongoing economic cost.

Availability and efficacy of containment and eradication measures

Pests such as fruit flies have a history of successful containment and eradication in many parts of the world (Myers and Hosking, 2002). Confidence in the possibility of successful eradication is a factor for consideration in any pest risk assessment. However, the financial and social costs of this type of programme impose a high level of precaution against the risk of introduction in regulated trade or any other means of entry. A range of techniques is applicable to containment and eradication including combined attractant and pesticide formulations, appropriate lures and traps for monitoring, inundative sterile male release and ability to control the programme by regulatory action.

Economic and social consequences

Production costs of commodities at risk from a newly introduced pest will increase in many instances. Additional monitoring and control measures would add significantly to costs of production and in some instances could preclude economically viable production of an at-risk commodity. Exceptions are where there are similar native or introduced pests with pest management regimes already in place which would be effective against the newly introduced species. Tephritid fruit flies are an example of this type of situation where it is not unusual for a number of species to be present in a country, making the increased difficulty of pest management less than it would be for a more unique pest. Further economic risks would arise from loss of access to markets where entry is dependant on a pest free status for production areas. At best this would incur the cost of measures such as a disinfestation treatment in which instance there would be a cessation of trade until access could be renegotiated and possible development costs of an acceptable disinfestation procedure, which might include ongoing intensive inspections with attendant costs. Other social and economic costs of a non-phytosanitary nature could arise with resultant interaction (Mumford, 2002).

5 Systems and Related Approaches to Pest Risk Management

A holistic pest management concept for meeting regulatory requirements termed a systems approach was outlined by Jang and Moffitt (1994) based on earlier proposals, such as Moffitt (1990) and Vail *et al.* (1993a), for pests of deciduous fruits, in particular, codling moth, *Cydia pomonella*. This concept of pest management for phytosanitary purposes recognized the contributions of the many routine production and operational procedures towards reducing likelihood of pest presence in a host at export and that the contributions to quarantine security could be cumulative. Systemic pest management can involve both natural and operational factors. If the pest reduction or risk mitigation potential of these procedures can be estimated reliably and optimized it is possible for them to replace some prescribed postharvest disinfestation measures currently required. Some of the factors can be quantitatively evaluated; others can be only subjectively assessed. ISPM 14 outlines the components of a systems approach as defined by the International Plant Protection Convention (IPPC, 2007). By their definition, a systems approach requires at least two or more components that are independent of each other.

Although in common usage, the term 'systems approach' is incomplete. It is a systems approach to pest management. A better term might be 'phytosanitary system', but in this chapter we mostly adhere to the IPPC defined term, 'systems approach' (IPPC 2007). Systems approaches should be composed of phytosanitary measures that can be implemented within the exporting country. The concept is capable of wider application than proposed initially and can range from production in regulated pest free or low pest locations to diverse multi-component arrangements that include phytosanitary treatments. The tradition of separating phytosanitary measures for reducing pest risk to acceptable levels into various non-overlapping categories seems to be breaking down as significant overlap exists among systems, pest free areas, areas of low pest prevalence, phytosanitary treatments and even some concepts of what is a non-host. Indeed, ensuring phytosanitary security has definitely embraced the integrated management philosophy of crop production, including pest management, food safety and secure passage through marketing channels, although the terminology may not have caught up with the practice.

Achievement of quarantine security through a systems approach can be compatible with an overall quality assurance aim and could even provide enhanced quality and economic benefits. There is an expected commercial incentive to produce the best practicable quality in a commodity and most of the quality benefits will flow from best practice in the production, handling, packing and marketing of the commodity. A phytosanitary system can replace or complement a prescribed phytosanitary treatment. The basis of prescribed postharvest treatment specifications is that infestation levels are not known but could represent an unacceptable quarantine risk and that a pest could be present at the highest incidence permitted by commodity marketability in a worst case scenario. Postharvest quarantine disinfestation is not a 'clean-up' procedure and must not be regarded as an alternative to good agricultural practice including, especially, good pest management in the production of a commodity, steps which could form the basis of a phytosanitary system.

Required Levels of Freedom from Pests

Planning and implementation of a phytosanitary measure based on a systems approach requires agreement on the acceptable pest infestation thresholds that need to be met. Acceptable thresholds differ between markets so systems approaches are notoriously difficult to standardize. Any threshold, even a nil tolerance, is only true at the level of inspection or surveillance employed and thus will have a statistical certainty that is always less than 100%. In practice, existing thresholds are rarely fully based on research data and most have been set empirically in the absence of full data. A widely used threshold is that of the USA originated by Baker (1939) that was set at 'probit 9' (i.e. 99.9968% mortality). Subsequently it was set at the usual confidence level (CL) employed for estimations of this kind, 0.95. For 99.9968% security this required no survivors from tests against approximately 93,600 pests (Couey and Chew, 1986; Liquido *et al.*, 1997; Cameron, 2002).

Landolt *et al.* (1984) estimated that the probability of establishment arising from introduction of one or more mating pairs from a shipment of 36,000 fruit, 99.9968% free of infestation was 0.027 for a mating pair. Assuming a 10% infestation rate before a treatment of 99.9968% efficacy, the probability of establishment would be 0.0032. At a lower initial infestation rate of 1%, this reduced to a probability of 0.0000657. In practice, the possibility of establishment of a new pest or disease in this way normally would be lower because of the expectedly low probability of finding a mate, hazards of survival in a new environment and finding a host suitable for oviposition. Acceptable commercial infestation rates for fresh produce for fruit flies and similar pests would be much lower because of the associated injury to the commodity. For many commodities the acceptable level is below the level of detection and this is achieved through routine production practices. So for some pests, postharvest disinfestation treatments, obligatorily prescribed, are clearly unnecessary. Other pests, particularly disease vectors or gravid adult females that have little direct effect on product quality, may need to be considered differently.

Baker *et al.* (1990) modelled fruit fly risk for New Zealand and proposed that the arrival of three live fruit fly larvae in imports on any 1 day from which a mating pair might develop after 3 weeks at any location constituted the maximum allowable pest limit to preclude establishment of tephritid fruit flies. They concluded that with good pest management in the field relatively few fruit in a consignment would be infested with fruit flies, but any infested fruit could contain numbers of larvae well above the maximum allowable limit. However, where a postharvest disinfestation treatment had been applied, even one not necessarily as efficacious as 'probit 9', it was unlikely that there would be more than one survivor in a single fruit. Sampling levels required to intercept infestations at their critical level vary according to the volume of host material accumulated on 1 day and recorded levels of infestation per unit of the commodity for each pest. This varied from 67 to 887 per shipment for the fruit fly pest species considered. Many of the species were so far out of their natural climatic range as to be incapable of establishment.

Jang (1996) used a sequential mortality model to show that natural mortality of fruit flies in 'Sharwil' avocados when supplemented by a mild 40°C hot water dip could achieve a quarantine security level of 99.9968% mortality. Subsequently, Mangan *et al.* (1997) modelled survival of Mexican fruit fly, *Anastrepha ludens*, in shipments of mangoes and citrus. They concluded that pest management measures were required during production to ensure that a postharvest disinfestation treatment of 99.9968% efficacy would meet a quarantine security requirement of < 1 mating pair of survivors in any shipment. For this pest in these commodities, it could be expected that standard orchard management and harvest conditions currently required by USDA-APHIS would be necessary to ensure ability to meet market standards for quality of fruit for consumption. These examples show how natural and routine production practices that reduce pest infestation can complement the efficacy of actions specifically oriented to quarantine security phytosanitation.

A systems operation manual for fruit fly hosts developed as a bilateral trade agreement between Australia and New Zealand incorporated many of the components applicable to a systems approach (Anon., 1993). It functioned successfully for a number of years before being superseded by development of a system with greater industry responsibility that reduced costly involvement of the regulatory organizations of the two countries. The original highly detailed system involved total coverage from production to inspection at export and again on arrival in New Zealand. Growers were registered, cultural practices for the minimization of field infestation prescribed and audited and trace-backs implemented if infested commodity units were found at either inspection. This enabled usage of postharvest disinfestation treatments of less than 'probit 9' (99.9968%) efficacy with consequent benefits to product quality.

Grains are traded on world markets in large tonnages and loaded into ships at rates (e.g. up to 3000 t/h) which make intensive sampling impracticable, even when automated. Australian grain was historically sampled under regulatory control at a rate of 1 l for each 40 t. Using a nil tolerance at this sampling frequency and obligatory disinfestation measures if a pest was found before loading from that grain stream could be resumed, there was anecdotal evidence that detections of pests on arrival at markets were reduced to < 2%. Two factors impinge on the efficiency of such a system for grains. One is the multiplication of undetected pests which occurs during transit, estimated to be a factor of 50 during a 6-week journey for grain at about 25°C and < 12% moisture content, the other the intensity of sampling at arrival.

Contributing Strategies to Pest Risk Management

A variety of measures may reduce pest risk, such as trapping, pesticide applications, physical barriers, harvest criteria and culling. When applied as an integrated system they may reduce the risk to acceptable levels. Table 5.1 lists steps in a systems approach to ship tomatoes from Morocco and the Western Sahara to the USA and the role of each step in reducing the risk of transporting Mediterranean fruit fly, *Ceratitis capitata.*

Pest free production areas

Known pest-free enclaves can occur naturally within the broad distribution range of most pests. These may consist of discrete ecologically isolated areas such as

Table 5.1. Phytosanitary systems approach for exporting tomatoes from Morocco and the Western Sahara to the USA (Source: after Hallman, 2007).

Step[a]	Role in risk management
Preharvest	
Limit production to three desert provinces	Sparse vegetation is poor habitat for fly
Grow tomatoes in 'fly-proof' greenhouses registered with national plant protection organization (NPPO)	Restricts fly from entering production areas
Maintain fly traps from 1 Oct. to 30 April	Detect fly populations
Capture of one fly in trap in greenhouse shuts it down until re-registration	Avoid picking infested tomatoes
Capture of flies outside greenhouses leads to bait sprays and placement of more traps	Prevent outside flies from entering greenhouses
Postharvest	
Export between 1 Dec. and 30 April	Period of low fly activity
Safeguard in 'fly-proof' covering in transit	Restrict fly access
Pack no later than 'pink' stage tomatoes	Pink tomatoes are at less risk than red ones of having fly (green tomatoes, although lower risk, are not profitable)
Pack tomatoes within 24 h of harvest	Reduce risk of infestation in packing house
Pack in 'fly-proof' boxes	Prevent infestation after packing

[a] Each step is considered to reduce the risk of transport of Mediterranean fruit fly, although quantification of the amount of reduction is generally lacking.

small or large islands or valleys in deserts (Fig. 5.1) or they may exist or may be created as production areas or even sites which can be maintained essentially pest free against endemic and migratory or dispersing infestations. Where these are suited to the production of a commodity it makes sense to utilize them as far as possible as a source for export shipments. According to ISPM 14, enclosed growing areas ranging from fixed structures to large netted areas can fulfil this requirement if linked with secure postharvest handling systems (IPPC, 2007).

Many countries accept the concept of 'pest free areas' (PFA) or 'area freedom' especially for fruit flies with host commodities involved in both domestic and internationally sourced trade (Malavasi *et al.*, 1994; Riherd *et al.*, 1994). The concept relies on initial negative pest incidence surveys, ongoing monitoring and eradication of any detected incipient invasions. It may be achieved also through prior eradication or ongoing suppression of a pest in areas in which low level persistence or invasions are possible. PFA status is used for tephritid fruit flies and codling moth but is applicable to other pests and pathogens where the preconditions can be met. Political boundaries may complicate regulation of PFA when the national boundary of the country seeking pest free status is not an ecological boundary and abuts another country in which the pest in question is known to occur.

Two IPPC Standards are primarily relevant: ISPM 04 and ISPM 10 (IPPC, 2007). ISPM 04 provides procedural guidelines for the recognition and maintenance of a PFA. It provides guidelines within which a country may act to

Fig. 5.1. Irrigated desert areas are ideal for establishment of insect-free areas to surmount quarantines. Asparagus fields in Peru are shown. Large yellow sticky traps help control some pests. Although border areas of other plants may augment natural biological control they may also provide havens for some quarantine pests.

determine whether PFA requirements might be met and the systems which need to be implemented to achieve this. Measures to establish and maintain the area pest free based on biology of the pest and relevant ecological characteristics are given in outline. These need to be considered within the context of the level of phytosanitary security required. This standard applies to three arbitrary types of existing PFA: (i) an entire country; (ii) a non-infested part of a country in which a limited infested area is present; or (iii) a non-infested area within a generally infested country. IPSM 10 outlines requirements for the establishment of pest free places of production and pest free production sites. These are smaller areas and even taking into account generous buffer zones would apply more to pests in which the adult dispersal ability or other infestation potential is limited. Potential examples are the mango seed weevil, *Sternochetus mangiferae*, the mango pulp weevil, *Sternochetus frigidus* and the sweetpotato weevil, *Cylas formicarius elegantulus*. Two other IPPC Standards, ISPM 06 (*Guidelines for Surveillance*) and ISPM 08 (*Determination of Pest Status in an Area*) are also relevant. A further ISPM, currently in preparation, will define areas of low pest prevalence for fruit flies.

The same fundamental principles apply to both systems in the accreditation of the PFA and the actions to maintain its status. In principle these are:

- Recognition of pest freedom.
- Accreditation.
- Maintenance of pest freedom.
- Identification of product.
- Maintenance of quarantine security throughout export pathway.

Recognition of pest free areas

Countries intending to import commodities which are known hosts of quarantine pests are likely to have promulgated requirements for the recognition of pest free areas of production. In conformity with the SPS Agreement (WTO, 2007a), USDA-APHIS defines a pest free area as a risk management option where the production area has been found completely free of a pest or is made free through specific actions and is protected from infestation or re-infestation. It cites areas in Mexico and Brazil as examples where this concept is used to ensure quarantine security against some fruit flies (APHIS, 2002b). Taiwan requires government authenticated surveys over time spans which vary from 1 year for most arthropod pest groups (moths, fruit flies, mites, thrips, whiteflies, scale insects and leaf miners) to 3 years for coleopterous pests and 5 years for fungi, bacteria and nematodes (COA, 2006).

Timing of exports to sensitive markets to seasons when the target pest is inactive or absent from the production area is a possibility. Most pests have a seasonal pattern of activity. When this is studied over a number of years it may be found that there are times during the production season of a commodity when a quarantine pest is inactive. Queensland fruit fly, *Bactrocera tryoni*, exhibits this type of behaviour being inactive in times of cooler temperatures in subtropical areas or areas of low humidity in dry tropics. Ideally, populations of all quarantine pests of a commodity should be monitored throughout a production system, either with traps or by sampling, which can be destructive or non-destructive. Seasonal freedom from risk of infestation can often be exploited as part of a systems approach. It requires organized marketing to ensure that sensitive markets are serviced during times of low risk if no additional pest management is to be undertaken (Armstrong, 1994).

Pests like the mango seed weevil, *S. mangiferae*, have patterns of dispersal such that new orchards can take many seasons before becoming infested. This is due to relatively sedentary habits which preclude unassisted dispersal beyond adjacent trees. Therefore if mangoes for export to markets which have prohibitions against this pest are identifiable as sourced from seed-weevil-free orchards beyond the range of natural migration of the pest from infested orchards, adequate quarantine security should be achievable without additional action. Destructive sampling, especially of windfall fruit, can establish freedom from infestation in an orchard or larger production area. Infested fruits can be identified frequently by a surface dimple which marks the oviposition point where newly hatched larvae entered developing fruit (Cunningham, 1991; I.C. Cunningham, personal communication). Establishment of site freedom from the pest requires a negative result from a statistically valid sample according to the standard required by the importing country.

The Government of India, through its Directorate of Plant Protection, Quarantine and Storage, has published a National Standard of Pest Management which defines requirements for establishment of PFA for both *S. mangiferae* and the mango pulp weevil, *S. frigidus* (Anon., 2005). A potential PFA is described as an area/municipality having a total of more than 400 trees, the minimum viable sampling unit. The standard sets out minimum numbers of both trees and fruit to be sampled and examined over a 2-year period and the stage of maturity. This standard is of interest because both pests are native to parts of the subcontinent but apparently are not as widespread as the host.

Eradication and verification for pest free areas

Pest free areas can be created by eradication followed by verification and subsequent exclusion of a pest from an area in accordance with ISPM 09 (IPPC, 2007), in particular, from areas of marginal suitability to the pest. This may require ongoing suppression such as with a sterile insect technique (SIT) methodology programme to suppress the establishment of populations of pests resulting from migration from adjacent harbourage areas or from incipient populations persisting at below the level of detection after eradication (Carey, 1991).

Other countries to make use of PFA created by eradication or suppression of a pest in defined areas include Japan (Oriental fruit fly freedom for Okinawa), Israel (Mediterranean fruit fly), Thailand (Oriental fruit fly) and Australia (Mediterranean fruit fly, Queensland fruit fly and codling moth). As an example, Australia established a fruit fly free area of approximately 400,000 km^2 known as the 'Tristate Fruit Fly Exclusion Zone' (Anon., 2001b) involving areas of the states of New South Wales, Victoria and South Australia for the production of citrus and other fruits intended for export. The pest freedom status applies specifically to Queensland fruit fly but the area is also monitored for incursions of Mediterranean fruit fly which is permanently established in Western Australia and is regularly detected and eradicated from nearby parts of South Australia (Fig. 5.2). The Tristate Exclusion Zone was naturally free of Queensland fruit fly in part because of the absence of native hosts. Irrigated fruit production enabled the fly to become established in years when seasonal conditions were favourable despite being outside of the long-term endemic limits (May, 1963). Area freedom is maintained by intensive monitoring with male lure traps, prohibition of travellers carrying host material which might be infested (Fig. 5.3) and control of incursions by SIT and field pest control with bait-sprays, all under regulatory supervision. This system, which avoids the cost of postharvest cold disinfestation treatment, has enabled trade to proceed relatively uninhibited with eight overseas countries, including New Zealand and the USA.

Maintenance of pest free areas

ISPM 04, 06 and 10 (IPPC, 2007) provide guidance for maintenance of PFA, covering preventive measures, exclusion measures, pest control measures and ad

Fig. 5.2. Mass rearing of Mediterranean fruit fly in Guatemala for production of insects for sterile release programmes against the pest worldwide. Capacity is for 3.5×10^9 flies/week; half that is currently being produced. Shown are towers of trays holding larvae in artificial diet as a rearing medium, held in a large controlled environment room.

Fig. 5.3. Unsupervised public involvement in maintenance of a pest free area for fruit fly in Australia. Regulatory enforcement roadblocks are randomly set up to encourage public compliance.

hoc inspections. Active monitoring under regulatory supervision will be necessary to assure authorities of the recipient area that the production area of origin is free of the pest and that it remains so throughout the production period. Monitoring can be done by trapping programmes, visual monitoring surveys by skilled operators or sampling, sampling usually being of a destructive nature. It is also necessary to have reliable procedures in place to avoid contamination of the export stream of the commodity with produce from non-certified pest free areas.

Surveillance trapping

Trapping is usually the preferred method of monitoring if applicable to a pest. The most highly developed trapping systems for phytosanitary purposes apply to fruit flies. There are numerous designs of traps (Fig. 5.4); most are variations of the old McPhail glass trap, itself a variation of traps used traditionally in Mexico to catch nuisance flies, or variations of a sticky surface trap, such as the Jackson trap.

Fig. 5.4. A range of designs of fly trap. Being able to adequately sample feral populations of quarantine pests is fundamental to the design and proper functioning of phytosanitary systems. Trap structure, colour, bait and placement may all affect capture.

Lures in these traps may catch one sex, such as male lure fruit fly traps which are specific to species groups. Other lures attract both sexes, for example McPhail fruit fly traps charged with ammonia-based lures (Cunningham, 1989b; Robacker and Landolt, 2002).

An action plan specifies trap density and thresholds for response to trapping finds. For some pest species, finding one individual may be judged to be sufficient reason to suspend accreditation and cause delimiting surveys and an eradication procedure to be commenced. However, for most pests depending on trap efficiency, it is recognized that a single pest individual may not represent a risk so an escalating series of actions needs to be agreed relative to the numbers and distribution of pests appearing in monitoring traps.

In the Australian 'Tristate' model (Anon., 2001b) male lure traps are located on a 400-m grid in residential areas and a 1-km grid in fruit production areas. Traps are estimated to have 8% effective catch efficiency and have been calibrated with respect to male/female frequency in the pest population. An outbreak is declared if five or more flies are found in two adjacent traps within a 14-day period, if one gravid female is found or if one larva is found in a fruit. An outbreak declaration results in quarantining of a defined area based on fruit fly migration behaviour and control measures involving cover spray applications of a systemic insecticide, or with bait-spray distribution, taking care not to disrupt natural control management of other pests. Alternatively, inundative releases of sterilized flies may be used if the species detected is Queensland fruit fly.

For other pests, especially codling moth, pheromones hold potential for detection and subsequent suppression. Traps may confine the trapped individuals, hold them using a sticky material or kill them by using a contact insecticide mixed with the lure. Ideally the trap design chosen should have the best possible catch efficiency, retain the trapped insects in a recognizable condition, enable determination of whether females are gravid and have practicable servicing intervals. Some traps rely on physical means alone, for example, light traps, air samplers, sticky surfaces, manual and automated grain sieves, pitfall traps, food residue samples or refuge traps such as those of corrugated cardboard or simple trunk banding.

Characteristics of insect monitoring attractants

Current procedures for the recognition and maintenance of PFA are greatly facilitated where attractants are available for the pests involved (Table 5.2). A recent review of these compounds and their usage by Robacker and Landolt (2002) recognized two main categories, food and pheromone based, and three primary roles, detection, delimitation and suppression. The first two roles involve trapping or sampling and the third, some form of pest population reduction which may involve an attractant.

Initially, pheromone-based control measures were expected to be developed readily for most pests, following characterization, synthesis and authentication of the compounds involved. It was believed that these would largely replace pesticides, providing environmentally acceptable systems of pest management free of pesticide residues. This wider expectation has not materialized but pheromones and other compounds which mimic pheromone responses, termed parapheromones (Cunningham, 1989a) have become available commercially for a number of species of Tephritidae. These pheromones (sensu lato) may be used in traps as lures, as mating disruption systems in field pest control or in quarantine eradication programmes combined with a pesticide in a trap or other distribution system as an annihilation system in quarantine eradication programmes, the latter being the most common usage.

The codling moth sex pheromone 'codlemone' is a chemical blend attractive to males. It was one of the earliest pheromone chemicals to become available commercially. Other pheromone blends have been developed but to date do not appear to be any more effective. Control by mass trapping with a pheromone lure proved ineffective in trials but lure-and-kill technology using droplet application on host foliage of the lure combined with a pesticide appears to be highly effective. Effective mating disruption with pheromones for codling moth and other tortricid moth pests such as the lightbrown apple moth, *Epiphyas postvittana*, depends on maintaining sustained release at appropriate concentrations but must avoid saturation of receptors of responder species which can occur at high

Table 5.2. Some uses of attractants in quarantine and other phytosanitary programmes (Source: after Bateman, 1982; Burkholder and Ma, 1985; Chambers, 1990; Robacker and Landolt, 2002).

Pest	Attractant	Usage
Lepidoptera		
Lymantria dispar, Gypsy moth	Sex pheromone	Detection, delimitation, suppression
Plutella xylostella, Diamondback moth	Sex pheromone	Detection, delimitation
Pectinophora gossypiella, Pink bollworm	Sex pheromone	Detection, delimitation, suppression
Grapholita molesta, Oriental fruit moth	Sex pheromone	Detection, delimitation, suppression
Cydia pomonella, Codling moth	Sex pheromone	Detection, delimitation, suppression
Epiphyas postvittana, Lightbrown apple moth	Sex pheromone	Detection, delimitation, suppression
Ephestia spp., Stored-product moths	Sex pheromone	Detection
Diptera		
Anastrepha ludens, Mexican fruit fly	Food attractant	Detection, delimitation of both sexes
Bactrocera cucurbitae, Melon fly	Parapheromone	Detection, delimitation, suppression of males
Bactrocera dorsalis (complex), Oriental fruit fly	Food attractant (protein hydrolysate)	Suppression of both sexes
	Parapheromone (methyl eugenol)	Detection, delimitation, suppression of males
Bactrocera tryoni, Queensland fruit fly	Food attractant (protein hydrolysate)	Detection, delimitation, suppression of both sexes
	Parapheromone (Cue-lure)	Detection, delimitation, suppression of males
Ceratitis capitata, Mediterranean fruit fly	Food attractant (protein hydrolysate)	Suppression of both sexes
	Parapheromone	Detection, delimitation, suppression of males
Coleoptera		
Anthonomus grandis, Boll weevil	Pheromone	Detection, delimitation, suppression
Popilla japonica, Japanese beetle	Pheromone, kairomone	Detection, suppression
Trogoderma granarium, Khapra beetle	Sex pheromone	Detection
Prostephanus truncatus, Larger grain borer	Aggregation pheromone	Detection

concentrations. Use of pheromones in this way has a greater possibility of success when it is a component of a multi-factor pest management programme (Robacker and Landolt, 2002).

More generally, stored-product pests also respond to pheromones and food-based attractants but according to order, rather than species, especially Coleoptera and Lepidoptera (Burkholder and Ma, 1985). These lures are either female sex pheromone lures or aggregation pheromones and are available for most important families of Lepidoptera and Coleoptera stored-product pests. Female-produced sex attractant pheromones are usually most effective for species with a short adult life including those belonging to Tortricidae, Pyralidae, Anobiidae and Dermestidae. Species with a typically long adult life including those belonging to the families Bostrychidae, Cucujidae and Tenebrionidae tend to respond more strongly to male-produced 'aggregation pheromones'. From a quarantine viewpoint the most important aggregation pheromones would be the lures for khapra beetle, *Trogoderma granarium*, other *Trogoderma* spp. and the larger grain borer, *Prostephanus truncatus*. However, because regulated phytosanitation is used more widely in trade in stored grains and grain derivatives, other sex hormones such as the multi-species moth hormone TDA and the Anobiidae beetle pheromone lures are relevant. There is some evidence that at least for stored-product pests, trap efficacy can be enhanced by the addition of a food-based lure (Chambers, 1990).

The parapheromone-based 'Trimedlure' has been the mainstay of surveillance and eradication programmes against Mediterranean fruit fly throughout the world for many decades. More recently, a defined food-based lure called 'Biolure' was developed and is expected to supersede Trimedlure despite the latter's high success rate and economic advantages (Robacker and Landolt, 2002). The male lure methyl eugenol has proved highly effective for the Oriental fruit fly, *Bactrocera dorsalis*, and together with 'Cue-lure' these two are capable of enabling surveillance trapping and suppression treatments against all but a few species of *Bactrocera* fruit flies of quarantine importance (Bateman, 1982). Food attractants based on various proteinaceous products have been widely used as baits in combination with insecticides in field control programmes for a range of fruit fly species but these materials have not been successful generally as trap attractants. Bateman and Morton (1981) made a comprehensive study of food-related attractants for *B. tryoni*. They concluded that gaseous ammonia is a powerful attractant for this species and that the attractancy of hydrolysed yeast fruit fly attractants is mainly due to the ammonia produced at the relatively low levels of pH characteristic of usual aqueous solutions of these materials. The attractancy was greatly enhanced at raised pH levels with the highest attractancy recorded for one hydrolysate at pH 8.5. Carbon dioxide was shown to have repellent action, as did very high concentrations of ammonia. This supports May (1963) who used ammonia-based lure in traps exclusively. Drew and Fay (1988) suggested that the attractancy of protein baits was more likely to be due to volatiles produced by bacterial growth in the proteinaceous attractant but this would not explain the long-standing consistency of catch with ammonia-based traps.

Other surveillance methods

Where the pest is an internal feeder, methods of monitoring can involve holding samples for pest emergence or destructive sampling such as by dissection of fruit taken at random from a tree or from among windfalls. Samples can also be taken from harvested fruit at the time of packing. For sedentary species such as scale insects non-destructive sampling as counts of scale numbers or non-infested fruit can be done provided that statistically valid sampling of all fruit on trees is possible. A biometrically authenticated sampling plan must be established. Pests of stored products are normally monitored by sampling storage bulks or during export loading but for some important phytosanitary pests, such as khapra beetle and other dermestids, a physical trap may be necessary (Pinniger, 1990).

Accreditation of Systems

This would normally be done as a bilateral agreement between regulatory authorities of the exporting country and those of the importing country, based on survey information supplied by the exporting country. Before agreement is reached, the importing country would make a technical assessment and be expected then to make its own technical inspection, unless pest free status was already in place with another importing country with similar certification requirements and technical standards.

Certification of product and whole of export path quarantine security

The primary means of certification of product from a PRA is the standard phytosanitary certificate. This would contain, as a minimum, declarations that, for a shipment or other required batch, the commodity was sourced from a PRA and maintained protected from infestation by the specified pest(s) along the handling pathway. To be fully effective, producer and handlers would need to be identifiable and audits conducted to maintain the integrity of the system. Where a breakdown is detected the system must be capable of effecting a rapid trace-back and rectification.

Field Operational Components

Some of the components of a phytosanitary system are preharvest and include field pest management, pest sampling and proper record maintenance during this period.

Optimization of field pest management for quarantine security

Good agricultural practices usually involve integrated pest management (IPM). The components of an IPM programme could include cultural measures, support or augmentation of beneficial organisms and 'soft' pesticides. It is based on economic injury levels that can be tolerated. One of the basic requirements of an IPM programme is establishment of the economic injury level (EIL; i.e. the level of pest injury at which economic loss occurs and the cost of pest management is justified) which can be tolerated, postharvest. For a commodity to be exported to a market with phytosanitary constraints this tolerance level can be factored into formulation of the IPM programme.

Apart from commodity varietal selection and timing of production discussed earlier, other components such as isolation of commodity production from alternate hosts or from other crops which may harbour a pest, even to the extent of being more favourable for pest development, should be taken into account. Tillage practices can be beneficial, for example hilling or irrigation of potatoes reduces the incidence of the potato moth, *Phthorimaea operculella*, in tubers at harvest. Fostering of beneficial insects as parasites or predators of a quarantine pest can facilitate pest incidence goals. Pests such as mealybugs (Hemiptera: Pseudococcidae) in citrus are normally controlled in an IPM programme by natural enemies including predatory ladybird beetles (Coleoptera: Coccinellidae), parasitic wasps (Hymenoptera), lacewings (Neuroptera: Chrysopidae) and insect fungal pathogens. These can keep population levels below EIL thresholds but where a lower level of incidence of this pest is required at harvest it can be achieved by early season supplementation of predators such as the ladybird beetle and predatory mites from insectary-reared populations. Similar management practices would be applicable to a number of pest groups including mites and scale insects.

Monitoring of pest incidence in crops

Surveillance, to ensure acceptable pest levels during production, is applicable to a wide range of commodities and especially for high-value horticultural products, mainly fruit, vegetables, nursery stock and cut flowers. For plant diseases, including viruses transmitted by arthropod vectors, it can be used for certification purposes or to trigger control treatments. As discussed above, pests such as tephritid fruit flies can be trapped using one or a combination of the synthetic or natural lures available. A problem with fruit fly lures used in this way is that the most conveniently formulated and efficient traps use longer life synthetic lures which attract only males and do not reliably detect dispersing gravid females seeking hosts for oviposition. Also, these lures are specific to species groups and more than one lure may be required to detect all fruit fly pest species that may be present at a location. Monitoring by host sampling is applicable to pests which are present from early in development of the commodity. An example is mango seed weevil discussed earlier in this chapter.

For cereals and other grains, there are fewer initial sources of pest infestation originating in the field. However, infestations of the bruchid *Callosobruchus*, *Acanthoscelides* and *Bruchus* spp. (Coleoptera: Bruchidae) in legume seeds can be of field origin. The maize weevil *Sitophilus zeamais* commonly infests maize in the field as do the nitidulid dried fruit beetles *Carpophilus* spp. and these infestations persist into storage. However, the main origin of infestations in cereal grain at harvest is from residues in harvesting machinery, storage and processing facilities and can be avoided by including insect sanitation into routine harvesting machinery maintenance programmes.

Trace-back facility

This ensures that pest interceptions during handling along the export pathway can be traced to a known producer, enabling identification of the source of any failure and, possibly, the reasons for its occurrence. It requires insistence on records of production operations including pest management, cultural operations and departures from normal routine procedures. Ideally, random audits are made a feature of such a system as well as following a breakdown of security. A positive feature is that producers can be briefed on best production practice with resultant benefits to production volume, quality and economic returns. Where a system of this nature is sustainable the benefits to quarantine security are substantial. It may prove that once an export channel becomes established some of the requirements or their frequency can be reduced, with economic benefits.

Research on seasonal incidence of pests and host phenology

An effective quarantine security system requires detailed, accurate and extensive knowledge of the pest and its interaction with the host commodity. This enables actions ranging from seasonal avoidance to timing of control measures. The widely researched Queensland fruit fly illustrates this principle well. In temperate climates this pest overwinters as the adult. Females develop eggs in spring, apparently related to rainfall patterns and gravid females then actively seek hosts. Produce which matures within the winter months can avoid infestation. However, in coastal areas and elsewhere where winters are not excessively cold oviposition may occur throughout the year although at a lower level in the cooler months. Overwintering numbers in cool dry areas are low and populations need a succession of hosts during the warmer months to reach potentially high infestation levels. Under these circumstances high-risk periods are from midsummer until autumn or early winter (May, 1963).

The phenological relationships between pests and hosts are varied and often complex. As well as those in the developmental biology of the pest there are many factors ranging from inherent host susceptibility through seed germination and flowering to maturity of fruit or vegetables. The interaction is often critical and research to provide basic knowledge on key aspects of these relationships will be amply repaid in the development of a quarantine security system. Vail *et al.* (1993b) undertook such a study on infestation rates of codling moth in walnuts. Despite the complications due to possible diapause in fourth instars, the stage present in stored in-shell walnuts, the risk of emergence and mating of normal residual numbers was shown to be low.

Operational: Harvest and Postharvest

During harvest and handling after harvest new factors come into play to reduce infestation risk and safeguard the harvested commodity from further infestation.

Avoidance of pest contamination at harvesting

Some pests of quarantine importance originate in other hosts and migrate to the export commodity at times as contaminating or hitchhiker pests. Where more favourable hosts are grown adjacent to the commodity to be exported the risk of cross-infestation is increased. Alternatively, mixed stands including plants unfavourable as hosts or those which actively repel the pest or act as trap crops can prove beneficial. Cultural measures, for example hilling of potatoes to protect tubers, are often applicable here. For cereals, other grains and oilseeds, sanitation practices with harvesting equipment will provide significant benefits.

Selection during harvest to avoid pest infestation

Commodities that are selected manually or electronically for colour or firmness during grading and packing can derive quarantine security benefits in this way. Produce can be selected for low infestation risk based on colour or ripeness for fruits including papaya, mangosteen and bananas. Damage can increase the risk of infestation by a quarantine pest. This may have occurred physically during operational procedures, from natural causes such as hailstones or secondarily as a result of injury from disease infections. An example is susceptibility of lychee fruit to some tephritid fruit flies where successful oviposition requires skin damage as oviposition into a sound aril has a low rate of success. Conditional non-host status in quarantine protocols can specify stage of maturity as well as absence of relevant injury.

Immediate cooling of harvested produce

Rapid reduction of field heat in produce at harvest is a well-recognized strategy for retention of harvest quality in perishable commodities, particularly fruit and cut flowers. Temperatures involved will vary with the tolerance of the commodity. Although the temperature reached after cooling of the produce immediately postharvest is unlikely to cause adequate mortality of pests to satisfy quarantine security standards alone, some reduction in pest numbers is likely if the temperature is maintained for the time involved in sorting and packing. Care must be exercised to avoid, if possible, cold acclimation of the pest which could influence effectiveness of a subsequent cold disinfestation treatment. Grains are likely to be harvested with considerable temperature variations due to time of day and other factors. Newly harvested grain bulks benefit from temperature equalization, which can be achieved through aeration or mixing during loading of storages. There can be instances where cooling of grain is disadvantageous if ambient temperatures are above pest optima.

Culling during the packing process

Fruits, vegetables and cut flowers are routinely subjected to culling during packing to remove items that are damaged, infested or otherwise unacceptable to a market (Fig. 5.5). Jang and Moffitt (1994) reviewed the efficacy of culling inter alia during packing and cited deciduous fruits susceptible to codling moth culled of infested fruits at an efficacy of 84%. Given that the incidence of fruits damaged by codling moth cited ranged from 0.2 to 34%, the process has potential for significant reduction in the level of infested fruits where numbers are at the lower end of the scale. Culling has useful potential for other pests causing external damage to commodities, even tephritid fruit flies in a range of fruits indicated by colour spots around oviposition punctures and feeding damage of thrips on flowers. Diversion of culls to non-sensitive markets can be employed where there is visual indication of heightened infestation risk such as with coloured papaya (Couey and Hayes, 1986) or those exhibiting a condition known as blossom end defect which increases susceptibility of green papaya to tephritid fruit flies in Hawaii (Zee *et al.*, 1989).

While most culling is a hand process done by operational staff, it can involve electronic equipment which can assess parameters such as colour and firmness.

Fig. 5.5. Certain steps in packing-house processing may form part of a phytosanitary system to reduce the risk of the final product carrying quarantine pests. Removal of older asparagus shoots in this case may reduce risk of infestation by lepidopterous eggs and early instars.

For seeds, mechanical cleaning and grading machinery has a long history of use. This machinery can remove insect pests which are external to the seeds with a high level of efficiency. Similar equipment is used in mills which are producing flours and related grain derivatives. Benefits are apparent when these products are to be exported to quarantine sensitive markets.

Cosmetic treatments

Simple cleaning of fruit by wet or dry brushing or pressure water sprays used to remove contaminants such as splashed soil and sooty mould can remove, also, associated hard and soft scale insects (Hemiptera: including Diaspididae, Margarodidae, Coccidae and other related pests including mealybugs). Walker *et al.* (1996) found that red scale, *Aonidiella aurantii*, could be removed from navel oranges with a high level of efficiency using a high-pressure water washer spray. Neven *et al.* (2006a) removed 90% of codling moth eggs from pome fruits with high-pressure washing.

The two commonly applied treatments to protect and enhance the appearance and shelf life of fruits are waxing and shrink wrapping. Saul and Seifert (1990) found that the addition of the insect growth regulator insecticide methoprene, formulated in the wax used to enhance fruit appearance, could achieve 99.9968% mortality of eggs and larvae of the tephritid fruit flies, *C. capitata*, *B. dorsalis* and *Bactrocera cucurbitae* in papaya in Hawaii, although the contribution of each component was unclear. Subsequently, Hallman *et al.* (1994) and Hallman (1996) found that partial control of Caribbean fruit fly, *Anastrepha suspensa*, was possible when fruits were coated with wax. A commercially available water-miscible wax coating containing amine fatty acid soap, waxes and food grade shellac reduced the number of immatures of the fruit fly in grapefruit and guava but was inadequate as a single treatment. However, this treatment controlled all exposed eggs nymphs and adults of Chilean false spider mite on cherimoya (Thompson, 1990; Unduragga and Lopez, 1992).

The coatings restrict gaseous exchange. Methyl cellulose and shellac, when applied to the surface of fruits, can cover eggs and trap surface pests and seal them, resulting in the death of the pest presumably by suffocation. Internal feeding pests are believed to be controlled by atmospheres that are modified within the coated fruits. Usually, carbon dioxide levels are increased and oxygen levels are decreased in pulp tissue (Sharp and Heather, 2002).

Treatments for other purposes can have an effect on a pest of quarantine importance. An example of this would be the effect of short-time hot water dipping for control of fruit surface infections, especially anthracnose, on the survival of fruit fly eggs in oviposition sites which are typically near the surface. Although the treatment could not be expected to be adequate as a stand-alone fruit fly treatment, a significant reduction in fruit fly egg viability would be probable at typical immersion times of up to 15 min at 51–55°C. Similarly, reduction of infection levels of pathogens occurs at fruit fly heat disinfestation regimes requiring a core temperature of 46–49°C (Jacobi *et al.*, 1994).

Conditions during pre-market storage

Perishable commodities are normally held at temperatures as low as the commodity will tolerate after grading and packaging. The time involved can range from hours or days to months. Such temperatures are usually disadvantageous to pests, ranging from suppression of activity to mortality at levels which meet the most stringent of quarantine security requirements and will usually reduce any postharvest infestation risk. Cut flowers and fruits or vegetables with low cold tolerance would probably be held at temperatures in the range 12–20°C for relatively short times with little benefit to quarantine security. However, citrus and pome fruits such as apples and pears, are typically cold stored for some months at 5°C or below with significant benefits to quarantine security. Longer storage times involving modified atmospheres with enhanced carbon dioxide levels have even more significant effects as discussed in Chapter 11.

Durable commodities may be held at low temperatures to protect quality and shelf life. Many seeds need to be kept as cool as possible to maintain viability. Nuts are a high value commodity which benefit from optimum low temperature storage conditions. However, grains harvested at times of the year when the temperatures are high can be stored beneficially at temperatures above the development thresholds of storage pests with consequent pest management benefits, especially where they are combined with modified atmospheres. Moisture content of each of these commodities has a bearing on pest susceptibility and subsequent pest development. Although moisture tends to equilibrate with ambient humidity, which may be advantageous or disadvantageous, this can be controlled to a greater or lesser extent by manipulating storage conditions.

Physical exclusion of pests, postharvest, can be achieved in a variety of ways. Insect sanitation measures in the vicinity of the packing shed or storage can decrease the incidence of opportunistic infestation by migrating pests. Some of these are high-risk pests of quarantine importance, others are incidental insect contaminants which result in general phytosanitary response if found in a product on arrival. Insect screening of ventilation openings of packing sheds together with flaps and air curtains on entrances and exits are widely used. So too are various forms of light attractants combined with a lethal system that may be chemical, electrical or physical. Packages can be made insect-proof by screening of vents in the case of perishables needing forced-air cooling or for small packages of durable commodities a laminated insect-proof package may be utilized. Stored grains held in hermetically sealed storages are equally protected if stored insect free or disinfested by some form of atmosphere modification.

Handling operations during storage

Grains can be handled during storage in a number of ways that can affect pest survival and development. A major factor in the movement of grain is impact mortality (Bailey, 1962). Several pest species are affected in this way including larvae of *Sitophilus* spp. and *Rhyzopertha dominica* within grains and larvae and pupae of *Oryzaephilus* spp. It can result from storage distribution systems in which the grain is 'thrown' at high speed but it can occur also when grain is conveyed to the top of a vertical storage and allowed to free fall a distance which permits adequate acceleration. Movement of grain will occur at intake to storage and subsequently in transfers or 'turning' to equalize temperature or moisture content gradients.

Pest monitoring during storage

Where it is appropriate, pest monitoring during storage provides a basis for control measures necessary to maintain the quality of the commodity. It is mainly relevant to grains, their derivatives and other durables such as bulk sugar. Where storages have been sealed and atmospheres possibly modified direct access may not be possible although storage parameters such as temperature and moisture content can be monitored using suitable remote read-out equipment. Storage in sealed packages has similar problems. However, wherever possible pest infestation of these commodities should be monitored and controlled if necessary to maintain product quality. Again significant phytosanitary benefits will ensue.

Specific disinfestation treatments

Many quarantine disinfestation treatments are a mandatory requirement for commodities traded into markets with prohibitions. These treatments are applied regardless of actual infestation levels. It is a tenet that they should not be relied upon as a part of routine pest management. However, where a commodity is traded without a mandatory requirement but must meet pest freedom requirements, a discretionary postharvest disinfestation treatment may be applied to ensure that the product meets market requirements. A discretionary treatment can be varied according to pest infestation risk with consequent benefits where the treatment carries the risk of injurious effects at levels required for extreme efficacies.

Handling and shipping

Handling and shipping temperatures, humidity and times all affect numbers of pests likely to be found on arrival at a sensitive market. These factors should, where possible, be managed to minimize the risk of a pest presence at entry inspections. Pest-free shipping facilities are a major component and measures should be taken to ensure that there is no residual infestation from a previous cargo. In-transit disinfestation treatments are most applicable to cold-shipped produce. The requirements involve suitably low temperatures, transit times in excess of the minimum for pest mortality and acceptable ways of maintaining and monitoring both time and temperature requirements. These are discussed in detail in Chapter 7. Routine operational practices, properly documented, can satisfy quarantine requirements without modification.

Sampling and Inspection

Few commodities would be exported without inspection if only to ensure that the product meets the specification required in the trade agreement. Rejection, possibly with remedial treatment, follows if the commodity fails the pre-export inspection. If a shipment of a commodity is sufficiently small, manual inspection of all units can include freedom from infestation as a requirement, given an ability on the part of inspection personnel to recognize infested units with a high level of certainty. Most export shipments are too large for this labour-intensive practice but sampling and the imposition of a nil or other tolerance level for pests of interest at inspection can ensure that phytosanitary requirements are met in one of two ways. For some pests a phytosanitary certificate carrying a declaration of 'no pests found' or 'not more than (a nominated number) of pests found' at a predetermined sampling level can be acceptable. Alternatively, a pre-export sampling with inspection (Fig. 5.6) and rejection of failed lots can be done at a level of intensity that will give an acceptable degree of certainty that the commodity will be found pest free at import. Such sampling can be destructive, including removal, or non-destructive which does not deplete the lot or shipment. The reliability of the inspection process in identifying an infested commodity must be commensurate with that of the importing country. Sampling for inspection is also used to monitor mandatory disinfestation treatments.

A sampling plan of sound statistical design is essential and must be followed rigidly. Ideally it should be drawn up by a specialist in biometrical statistics and will take into account the effect of destructive sampling. A sample is an independent unit of a shipment or subshipment of commodity which enables

Fig. 5.6. Sampling a grain stream for insect infestation. The sample will be sieved to separate insects from the grain. Internal stages cannot be determined. Automated sampling is a feature of some modern terminals but pest identification should be done by inspectors. (Photograph by M. Bengston)

inferences to be drawn about the lot from which it is taken if selected in a statistically valid way. Sample unit numbers determine the statistical confidence level of the inferences, provided that all sampling theory assumptions are met. For commodities in which individual units can be infested these can represent sample units. Other commodities such as grain need to be sampled as measured units that can be reliably assessed (e.g. 1 l). The lot being sampled should be discrete with respect to origin and otherwise as homogeneous as possible. Sampling should be done in a statistically random manner, although in practice it is frequently possible only to select samples without conscious bias. Trained operators are essential to ensure this aspect of sampling, as well as the consistent assessment samples for pest infestation and identity.

Sampling of consignments for visual phytosanitary inspection

Sampling and inspection in which each independent sample unit is assessed as non-infested or infested is analogous to the experimental procedure described in Chapter 6 where treated insects are assessed as dead or alive, each representing an experimental unit (Couey and Chew, 1986). Tables and software (Liquido *et al.*, 1997; Cameron, 2002) are available giving the numbers of sample units that need to be taken from a lot and examined to give a required level of statistical certainty that a level of security has been achieved. Thus, for example, 598 sample units found to be pest free would give an assurance that less than 0.05% of a lot is infested at a 95% CL but 919 samples would be required to meet this level of risk at the 99% CL if sensitivity is 100%. Some software sources also allow calculation of the security achieved if there are survivors or if the sensitivity is not 100%. Couey and Chew (1986) provide the following equation to this condition:

$$C = 1 - (1 - p_u)^n$$

where C is the confidence level, n the number of samples or experimental units and p_u the upper confidence interval value. Where there are one or more survivors detected the calculation of n becomes more complex involving approximation with a Poisson distribution. This involves large numbers of n usually impracticable to examine in a manual inspection. In general, the sample size to be examined is that designed to achieve the required level of security at a given confidence level for a population of stated size, assuming no survivors are found. However, countries may decline to accept produce in which any infested units have been found at inspection, regardless of the calculated risk, because of public perception.

Conclusions

A multi-component systems approach to quarantine security shows great potential for expanded use in many areas. It presents more difficulties in terms of acceptance than a single treatment conferring known quarantine security but will have fewer negative effects on product quality and could open the way to trade not currently possible. The main difficulties are in quantification of risk arising from each component and the integration of these to provide a reliable estimate of end point risk. Gathering the data will be long term and expensive and needs to be offset by the probability of substantial trade in the commodity. However, as knowledge of each of the components is accumulated there will often be related benefits in pest management and production of the commodity generally.

6 Development of Postharvest Phytosanitary Disinfestation Treatments

The Agreement on the Application of Sanitary and Phytosanitary Measures (SPS Agreement) arising out of the Uruguay Round of negotiations of the General Agreement on Tariffs and Trade (WTO, 2007a) requires that phytosanitary measures be technically and scientifically justified. Phytosanitary treatments can reduce the risk of moving pests in trade, thereby surmounting trade barriers based on phytosanitary concerns. Treatments should have technically proven levels of efficacy whether used as single treatments, components of multiple treatments, or in combination with other pest mitigation measures as part of a phytosanitary systems approach. Phytosanitary treatments kill, inactivate or remove pests from regulated articles or render the pests incapable of completing development or reproduction (IPPC, 2007). If the pest itself is a plant, such as seeds or other propagative material, the objective of a phytosanitary treatment may be devitalization, rendering plants or plant products incapable of germination, growth or further reproduction.

The information required of an exporting country to demonstrate efficacy of a phytosanitary treatment must be scientifically sound. Information on the pest should include taxonomic identity, most tolerant stage of the pest to the treatment, source of the pest population, rearing conditions, method of commodity infestation and evaluation method, including size of large-scale trials and survivors, if any. Information on the regulated article if applicable should be the scientific name, cultivar, if available, and all known attributes such as origin, stage of development, weight, shape and market quality. Information on the research methodology should include a description of the research facility and equipment, experimental conditions (temperature, relative humidity, diurnal cycle), method of dosimetry and its calibration, critical treatment parameters (dose, exposure method and time, temperature and any other factors that may affect efficacy) and precise description of what constitutes treatment efficacy and under what conditions.

Research should commence with dose-response bioassays of the pest and the commodity. These provide data that may be analysed statistically to estimate a dose required to achieve a treatment efficacy at the desired level of quarantine security. Tolerance of the commodity to the treatment may also be investigated. Because the estimated requirement to achieve quarantine security is effectively 100% control, the statistical confidence placed on this estimate is very low unless sample size is extremely large.

A number of variables must be considered for the development of any phytosanitary treatment. In general, a large number of pests of the stage most tolerant to the treatment that would normally be present in shipped commodities must be treated. The pests should be treated at one specific dose or precisely defined, monitored and controlled set of dose conditions that represent commercial practice with no survivors or only a very few predetermined to be acceptable. In addition, appropriate controls must be used throughout the research and these controls must respond to whatever end point is being measured within normal ranges. The presence of one survivor among tens of thousands treated can sometimes invalidate a treatment; therefore, special care must be taken to keep treated and control organisms separate and prevent re-infestation of the treated commodity, always a risk in phytosanitary research. A false negative (the treatment was not observed to be efficacious when it was) will lead to more stringent treatment than necessary, resulting in inefficient use of resources and possible damage to the commodity. A false positive (the treatment was observed to be efficacious when it was not) may increase the risk of an invasive species becoming established in a new area.

Types of Treatments

There are many factors to consider when deciding what treatment to attempt to surmount a given quarantine barrier. The major ones are dealt with in six individual chapters following this chapter. Chapter 13 discusses several treatments of limited research interest or commercial application. General comparisons of a number of treatments are given in Table 6.1. More detailed considerations are given in each relevant chapter.

Commodity Characteristics

Plant-based commodities may be broadly categorized as perishable or durable. Perishable commodities include fruits, vegetables, cut flowers, green foliage, live ornamental plants and plant parts for propagation. Durable commodities include harvested cereal grains, seeds, dried foliage, dried herbs, spices, logs, lumber, cotton bales and dried tobacco. Perishable commodities have a short shelf life and are generally more susceptible to handling and phytosanitary treatment injury than durable ones. Many durable commodities do not consist of plant material but may harbour incidental pests or contain packing materials that are hosts of pests. Examples of these are ceramic tiles and brass objects, both resulting in many interceptions of quarantine pests.

Table 6.1. Comparison of phytosanitary treatments for various properties (Source: after Hallman, 2007).

Treatment	Commodity tolerance	Cost	Speed	Logistics	Acceptance by organic industry
Cold	Moderate	Low	Very slow	Easy	Yes
Heated air	Moderate	Moderate	Moderate	Moderate	Yes
Hot water immersion	Moderate	Low	Fast	Moderate	Yes
Radiofrequency heating	Moderate	Moderate	Fast	Moderate	Yes
Ionizing irradiation	High	Moderate	Fast	Moderate	No
Methyl bromide fumigation	Moderate	Low	Fast	Easy	No
Sulfuryl fluoride fumigation	Low	Low	Fast	Easy	No
Phosphine fumigation, solid formulations	Low	Low	Moderate	Easy	No
Phosphine, gaseous formulations	Moderately low?	Low	Moderate	Easy	No
Modified atmosphere (MA)	Moderate	Moderate	Slow	Moderate	Yes
MA/heated air	Moderate	Moderate	Fast	Moderate	Yes
Non-gaseous pesticides	High	Low	Fast	Easy	No

Taxonomic identity of plant hosts can become confused where hybridization has been induced. For most commercial crop plants long-term breeding programmes and more recently genetic engineering has produced recognizable cultivars. Varietal differences can affect host susceptibility or treatment efficacy. For example, pineapples that are ⩾ 50% 'Smooth Cayenne' pedigree are not considered by the USA to be hosts of tephritid fruit flies, and thus do not require a phytosanitary measure to enter that country if fruit flies are the only concern (APHIS, 2007). Other genotypes of pineapples may require a phytosanitary measure.

Although cultivar differences are usually insignificant they can provide grounds for disputes about treatment efficacy in the absence of evidence to the contrary. For example, Japan held the view that different cultivars of fruits could yield different efficacy responses to methyl bromide fumigation for codling moth, *Cydia pomonella*, and demanded complete research support before any new cultivar would be imported using methyl bromide fumigation as a disinfestation treatment. This was the first phytosanitary trade barrier case taken to a dispute settlement panel of the World Trade Organization (WTO). The findings from the ensuing deliberation were: the varietal testing demand was not based on sound science; Japan did not attempt to obtain additional information to review its more restrictive actions in a reasonable time frame as it should have; the panel was unable to rule on a proposal by the USA to base methyl bromide fumigation treatments on a product-by-product basis for lack of sufficient scientific evidence; and the varietal testing requirement should have been a transparent part of Japan's phytosanitary requirements before the issue was raised by Japan to the USA (WTO, 2007c). In 2001 a mutually acceptable agreement was reached by both parties and further WTO action ceased.

New Zealand placed similar restrictions on dimethoate-treated tomatoes imported from Australia (Anon., 1993). For many years, only those varieties which had been individually tested at an efficacy level of 99.99% at a confidence limit (CL) of 95% against *Bactrocera tryoni* were permitted entry despite negligible differences between cultivars. The issue was eventually resolved so that all table tomatoes were accepted without further testing.

A recognized specialist should confirm the identity of any cultivar used in disinfestation trials. For plant species archiving of test material is usually unnecessary but inclusion of appropriate photographic records when reporting results could be advantageous. Correct definition of the commodity is particularly important where one or more species of commercial plants can be known under a single common name. For example peppers can be 'sweet', *Capsicum annuum* (which may be green, red or yellow in colour) or 'red', *Capsicum frutescens*. Similarly, within a commodity group some cultivars can have different taxonomic origins. For example, of the commercially marketed easy-peel citrus (mandarin) varieties, 'Imperial' is taxonomically *Citrus reticulata*, a true mandarin or tangerine, 'Ellendale' and 'Murcott' are *C. reticulata* × *Citrus sinensis* (tangors), while 'Minneola' is *Citrus paradisi* × *C. reticulata* (tangelo).

The host substrate used to develop the treatment must be completely and properly characterized. Size and shape are two of the most obvious variables affecting efficacy of heat treatments. Mango hot-water immersion times depend on the weight and shape of the fruit; lighter and flatter varieties require less time than heavier, rounder ones (APHIS, 2007). Varietal morphology can lead to differences between commodities such as in adsorption of fumigants through differing skin texture or 'furriness'. Treatments with pesticides having systemic action may also be affected leading to residues in excess of approved Maximum Residue Limits (MRL). The broadest possible applicability should be negotiated with the importing country from the outset and common practice is to relate treatments to commodity species or major groupings rather than cultivars, if possible.

Pest Identity

It is of utmost importance to define a pest taxonomically and to ensure that the appropriate species is used in disinfestation tests. The taxonomic identity of the pest should be validated and agreed upon with the proper authorities of the importing country as part of a pest risk analysis (PRA; as explained in Chapter 4) before developing detailed research plans for disinfestation treatments. Voucher specimens must be retained in a secure location, such as a tenure museum, able to be referenced following any reorganization of relevant taxons or if there are other questions regarding identity of the research organisms.

Similarities in response may be usual across species in a genus and are even likely among genera in a family. For example, Hallman and Loaharanu (2002) proposed that a low radiation dose of 70 Gy could serve as a phytosanitary treatment for all *Anastrepha* spp. fruit flies because research with several species yielded quite homogenous responses. They further proposed a dose of 150 Gy for all genera in the family Tephritidae.

Taxonomy of most pest species is subject to ongoing review and despite the facility in systematics for the conservation of names based on long usage, it is not unusual for nomenclature of a pest to be changed one or more times over a period of a few years. Also, a species entity may be divided into two or more species so that subsequently doubts may be held about the identity of the original test population. For example the division of what was considered the Oriental fruit fly, *Bactrocera dorsalis*, into 41 species (Drew and Hancock, 1994) cast into doubt the validity of phytosanitary irradiation research performed in several Asian countries (Hallman, 1999). That doubt could have been avoided had voucher specimens been available.

Where new species are geographic races, morphological differentiation may not be possible and DNA studies may be required. Usually, the original experimental population is the taxon of greatest pest potential but it could have been what is later suspected to be a mix of different species. Modern taxonomic practice grants species status to groups that may interbreed readily in the laboratory but not in the field. When these species occur in the same location and are morphologically very similar, experimental populations sourced from the field or supplemented by field collections can prove to be taxonomically suspect unless great care is taken with identification and archiving of a series of specimens.

Test Populations of Pests

Ideally, field populations of quarantine pests in naturally infested commodities from prospective exporting areas would be used to develop phytosanitary treatments. The advantages are that these organisms are exactly the ones that have resulted in the quarantine and they are in their natural relationship with the commodities to be exported. The risk that the choice of pest individuals, commodities and infestation and holding conditions will result in the treatment being erroneous is minimal. However, using feral infestations is usually not possible for several reasons. Quarantine pests in the natural setting are frequently not abundant enough to work with; often the naturally infested commodity is not in a good condition for treatment; it is difficult to control infestation levels and stages in a way that would yield good progress in the research; and the pests may be infested with diseases and parasitoids, making mortality counts unreliable. Nevertheless, for some pests such as the mango weevils, *Sternochetus* spp., rearing in the laboratory may not be possible and using naturally infested commodities may be the only viable method.

Testing of host susceptibility and the efficacy of proposed phytosanitary treatments frequently requires the establishment and maintenance of laboratory populations of pest species for one or many generations. Laboratory populations can also be held for the purposes of inundative release of sterile individuals to eradicate or suppress a pest in an area of phytosanitary importance (sterile insect technique, SIT). Methodology may differ according to these purposes but in general, conditions applicable to one are relevant to the other. Exceptions would be populations specifically modified for SIT, such as temperature-sensitive lethal Mediterranean fruit flies, *Ceratitis capitata* (Franz *et al*., 1994). Genetically modified organisms or those bred by conventional means should not be used for phytosanitary research unless it is clearly demonstrated that their response to the treatment does not differ significantly from feral organisms.

Naturally occurring populations of pests can be expected to exhibit variation, which may be either environmental or genetic in origin and these influences can interact to produce a genotype × environment response to treatment. Hence there is likely to be a critical sample size for representative genetic diversity. Diversity or heterogeneity is important also for genetic fitness; consequently, genetic diversity is desirable in a representative test population of a pest. Dispersal of a pest within its area of occurrence has elements of both time and space and these must be taken into account when establishing and maintaining a laboratory population, so samples should be collected from differing localities and at differing times of the season of occurrence (Bartlett, 1985). Heterogeneity leads to wide experimental limits in response to many treatments and can mask small differences such as comparative responses between stages.

Laboratory cultures should be sourced widely from field-infested host material. Collections should be made throughout the potential export range of the host and the pest species. Cohorts of at least 200 mated females should be collected to ensure an acceptable probability of sampling the genetic variation in a contiguous population. Care should be taken not to restrict the sample to pieces of adjacent host material as these could be infested by progeny of a single female. Where there is a possibility of more than one pest species in a host fruit or where parasitism is suspected, it may be necessary to isolate each pupa or final instar nymph until the adult emerges. Viable unmated adults of the same species can then be grouped for breeding subsequent generations.

Freedom from pathogens and good physiological vigour are two essentials for successful rearing of test populations. Great care must be taken to ensure that population samples do not carry diseases present in wild populations. This is best achieved by rearing field-collected samples in quarantined conditions until it can be ascertained that they are disease free; it applies to both initial establishment and to supplementation. Physiological vigour may be ensured through favourable diet and environmental conditions during rearing.

Cultures should be managed so that pests are available at an acceptable level of fecundity at the time of the trial and in adequate numbers to infest the experimental commodity to the level required. Overall vigour can be monitored in each generation or otherwise as required through characteristics such as survival percentages for each stage, fecundity, fertility, weight at a key stage and longevity of adults. As a guide to the vigour of culture populations, the following parameters could be monitored for each generation:

- Survival from egg to pupa (or last stage nymph).
- Adult numbers as a proportion of pupae or last nymphs.
- Mean pupal or last nymphal weight appropriate for the species.
- Development times of cohorts, egg to pupa and adult.
- Sex ratio of enclosed adults.
- Fecundity.
- Flight ability where applicable.

It is usual practice to supplement the genetic diversity of long-term laboratory cultures by the addition of field-collected individuals, adults or juveniles, on a regular basis, for example annually.

Regimens used during rearing will create inevitable genetic selection pressures on a laboratory population. For example, Economopoulos and Loukas (1982) recorded genetic change in olive fruit fly, *Dacus oleae*, after only four generations of rearing on a semi-artificial diet. If susceptibility is increased in culture the treatment may fail when applied to wild population pests. If the opposite happens and susceptibility decreases in culture, risk of establishment of the pest will not increase, but a treatment may be more stringent than necessary. Corcoran (2002) found that laboratory-reared Queensland fruit fly, *B. tryoni*, were more tolerant to heat disinfestation treatments than wild individuals, which may have been due to genotype, phenotype (favourable rearing conditions) and/or the interaction of the two. In another example, Hallman (1994b) found that the rearing temperature of Caribbean fruit fly, *Anastrepha suspensa*, affected susceptibility to hot but not cold temperatures used in phytosanitary treatments. Third instars reared at a constant 30°C were significantly more tolerant of hot water immersion than those reared at lower temperatures.

Consistency in the use of substrate, nutrients, supplements, lifespan, rearing temperatures and humidity are all important in rearing test insects which need to respond as consistently as possible to test regimens. At the same time, test populations must be representative of wild populations encountered in commercial operations. Literature on rearing insects is extensive. Singh and Moore (1985) prescribed laboratory-rearing requirements for 86 insect species from ten orders and Winks (1983) listed requirements for 110 species of stored-products pests from six orders. Anderson and Leppla (1992) emphasize the rearing of insects for research purposes as opposed to mass production. Cohen (2004) and Siddiqui and Dey (2002) provide recent volumes on insect diets.

Rearing, Holding and Treatment Facilities

Various supporting services and equipment may be needed to generate the information required for a proposal for a disinfestation treatment or a system to ensure phytosanitary quarantine security:

- Rearing facilities for pest colonies.
- Holding facilities for infested commodities.
- Security against entry or escape of pests to and from infested commodities and between treated and control (untreated) commodity samples.
- Adequate cages and containers for pests and treated commodities and storage facilities for rearing materials and holding commodities to be used in research.
- Laboratory equipment for preparation of culture media including sterilization, counting, weighing, microscopic examination, recording and computation of results.
- Facilities capable of treating experimental samples at ranges of doses required.
- Laboratory equipment for dosimetry measurement.

Day length, temperature and humidity should be controllable, although facilities that are open to ambient conditions may be preferable when those conditions adequately represent ambient conditions in the commodity production areas or where the pests need to be reared on living plants. Temperature and humidity should be recorded continuously so that trace-back is possible if apparent anomalies in results occur. Variations in temperature, and to a lesser extent humidity, will influence time of development and hence availability of test insect cohorts at set times and homogeneity of developmental stages of test samples. Controlled lighting may be needed to simulate an optimal day length or dusk lighting, which can be a mating requirement, but natural lighting can be a requirement for some species. Alarm systems for equipment failure should be fitted and operational throughout the conduct of an experiment, as should monitoring and recording systems such as charts or data-loggers. Some pests which cause breakdown of the host tissue, such as fruit flies, may require that drainage slits be cut in fruit to avoid drowning of larvae and ventilation to enable ready escape of carbon dioxide from pest metabolism and host tissue breakdown. It is important to maintain genetic variability in the cohorts, both pest and commodities, used in the large-scale confirmatory tests to capture the bulk of natural variability among the samples and the variability in susceptibility to the treatment that this may entail.

Host Material

Testing treated samples requires commodity free of prior residual effects of any pesticides applied during production. Even low levels of pesticide residue may prevent development of the test insects in commodities. For example, organically grown apples were used in tests with apple maggot, *Rhagoletis pomonella*, because apples from sources that used pesticides did not yield infestations at times nor did ovipositing adults survive long (Hallman, 2004b). Host material can be monitored for the presence of pesticide residues by chemical analysis such as gas chromatography or, preferably, by a suitable bioassay with a susceptible strain of the pest. This applies also to ingredients of culture media. It is easily possible to lose an entire laboratory population through low level contamination of host or culture material.

Commodities selected for research should represent the range of the potential export production area regarding cultivar, size, quality and growth stage. A hot-water immersion treatment for Hawaiian papayas against tephritid fruit flies failed a couple of years after development (Zee *et al.*, 1989) because the treatment was designed to kill only eggs and first instars and relied on early stage of papaya ripeness to ensure that later fruit fly stages would not be present. Subsequently third instars were found in treated papayas and it was discovered that some papayas had a 'blossom end defect' (a small hole to the seed cavity of the papaya) in which fruit flies could oviposit earlier than they would in a papaya without this hole. The large-scale confirmatory testing to support this treatment was done with papayas from only one orchard (Couey and Hayes, 1986). Perhaps a broader selection of papayas would have led to the detection of the open blossom end problem before the research was concluded.

In some cases surrogate host material may be used to develop a treatment. Hallman (2004a) developed an irradiation treatment against oriental fruit moth, *Grapholita molesta*, using an artificial diet after determining that radiosusceptibility of the pest reared in a natural host (apples) was not significantly different to radiosusceptibility of it in the diet. It was easier and less costly to conduct the large-scale testing in the diet.

Criteria for Efficacy

Acceptable outcomes of phytosanitary treatments can range from immediate mortality of the pest to prevention of its establishment in a new area through sterility or other physiological inability to reproduce successfully. Immediate mortality of the stage treated is normally simple to determine and record but mortality in a subsequent stage requires that the treated samples be held for a time. Determination of sterility or other inability to reproduce requires both holding under favourable conditions until the next generation and trials to determine whether any reproduction is successful. The criterion used for efficacy of a treatment is highly important as it may lead to problems at subsequent sampling and inspection of a commodity when live pests are found after a valid treatment. If these are juveniles and the treatment has been shown to prevent development to fertile adults, it should not constitute a basis for rejection of a shipment on quarantine grounds provided that the treatment can be validated. Similarly, if adults are found and an approved treatment known to result in adults incapable of breeding is properly applied and all other safeguards preventing re-infestation or contamination are in effect, a shipment should be passed with confidence.

Efficacy should be defined explicitly and the threshold for failure of the treatment established. It is especially important for phytosanitary inspection that inspectors understand what constitutes an effective treatment and what constitutes treatment failure. In practice, however, there may be a gap between what the scientific information supports and what regulators count as effective or ineffective treatment (Fig. 6.1). For example, the research supporting a methyl bromide fumigation treatment against Mexican fruit fly, *Anastrepha ludens*, allowed for pupariation in its definition of efficacy (Williamson *et al.*, 1986). However, regulators tend to consider treatment a failure if any live larvae are discovered.

A number of publications discuss level of security required for phytosanitation. Landolt *et al.* (1984) proposed that the level of security be based on the probability of a mating pair (in the case of Tephritidae) occurring in a commodity shipment. Baker *et al.* (1990) estimated the commodity infestation level that would prevent even a mating pair from surviving given infestation rate and treatment efficacy, based on an amount of infested commodity entering an area susceptible to pest invasion. Vail *et al.* (1993a) showed that infestations of codling moth in nectarines and cherries in California were very low and that phytosanitary measures required by some countries greatly exceeded what the authors considered to be reasonable phytosecurity needs. Mangan *et al.* (1997)

Fig. 6.1. Phytosanitary researchers may employ techniques to deal with large numbers of insects that may differ in key ways from how regulatory inspectors interpret their findings. For example, stacking infested, treated fruit in towers where larvae emerge and fall into a soil-like substrate below for capture and further development has been used in many studies with tephritid fruit flies. Flies remaining in the fruit may not be counted and the criterion for efficacy is beyond finding live larvae in fruit, whereas inspectors usually count live larvae in fruit as failures of treatment, except for ionizing irradiation.

applied the concepts of Baker *et al.* (1990) to Mexican fruit fly in mangoes and citrus fruit from Mexico and concluded that if fly abatement programmes in groves failed, levels of surviving flies could exceed the accepted level of security even if a phytosanitary treatment was applied. Follett and McQuate (2001) discuss pest risk associated with developing phytosanitary treatments for commodities that are poor hosts of quarantine pests.

Determination of Pest Stages to Treat

For any quarantine disinfestation treatment, it is important to determine the most tolerant life stage(s) of the pest occurring on or in the commodity for export. This aspect is discussed in detail in chapters specific to treatment technologies. The stage to be used for determination of the operational treatment dose needs to be agreed by the importing country or trials to demonstrate its status will be necessary. Where a treatment does not kill all individual pests immediately, or where this cannot be verified, such as for internal pests, experimentally treated samples must be held under conditions favourable for pest development until the number of survivors can be determined. Parallel non-treated (control) samples for the estimation of initial infestation levels in treated samples must be held under similarly favourable conditions until all of the treated individuals are found by visual inspection or other means to have emerged or died. If a treatment causes only sterility it must be tested by controlled matings of F_1. Development and reproduction in untreated controls must be normal.

Tolerance comparisons should be made in the upper range of control, such as 99% (a level estimated by most statistical software) and not at the 50% level. Because efficacy must be demonstrated to a high degree of precision a higher level of error should be accepted when analysing data to determine most tolerant stage. It is recommended to use the stage in confirmatory tests whose upper 95% CL is greater when the means are roughly the same.

Influence of Treatment Technology

The disinfestation technology to be used will influence the design and conduct of the experiments. Physical treatments such as cold and heat often have a pre-treatment time component while the commodity reaches the treatment conditions. For heat, the warm-up time will usually be relatively short but for cold the pre-cooling time could involve many hours, even days. These times, if not consistent, can lead to experimental variability because pre-and post-treatment cold exposure and pre- and post-treatment residual heating contribute to eventual treatment mortality. For irradiation, treatment is immediate and treatment time normally short. Treatments involving fumigants also take effect almost immediately, although there will be a gassing-up period as the existing atmosphere is displaced or modified and decay of gas concentration will occur during the treatment period due to sorption and leakage. Pesticide dips or re-circulated flood sprays may be subject to lowering of concentration through loss of the active ingredient on previously treated product, known as 'stripping' (Noble, 1983).

Consideration must be given to how a treatment will be applied on a commercial scale and how the research phase may differ (Fig. 6.2) to avoid conducting research that may be irrelevant to the commercial treatment situation. For example, cooling in a small chamber can be done faster than cooling in a large commercial facility. Slower cooling in the commercial facility can be expected to enhance the lethal effect compared with faster cooling in an experimental unit but it can result in reduced mortality if cold acclimation occurs (Iwata *et al.*, 1992). Gould and Hennessey (1997) found that rapid cooling (45 min) of carambolas infested with Caribbean fruit fly to 1.1°C reduced the expected time to achieve quarantine security by one-third compared with cooling that took 24 h. As the size of heated-air units increases maintaining treatment uniformity becomes more difficult and post-treatment cooling time may contribute unevenly to overall mortality.

For each treatment technology there will be specific factors to be considered that relate to the way in which dosimetry is done and experimental samples of

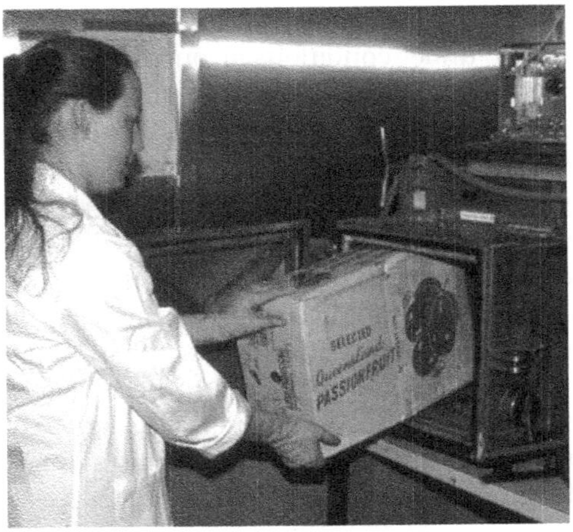

Fig. 6.2. Small-scale research fumigation of test fruit. Potential differences in efficacy and commodity quality between small, research-scale and large, commercial-scale treatments must be addressed before phytosanitary treatments can be offered to industry. (Photograph by R. Corcoran)

pests and hosts are handled. Where host sample units are small, as in most initial experimental stages (Fig. 6.3), precision is of extreme importance in the administration of doses of the treatment. The number of sample units of the infested host commodity will depend in part on the capacity of the experimental treatment unit although buffer samples may be necessary to ensure that the test is done at normal chamber capacity. In most instances this will be smaller than commercial treatment facilities. However, the treatment administered must be comparable with that of a commercial facility (Fig. 6.4). This may require construction of experimental treatment units with special receptacles if the commodity cannot be treated in standard packages such as those used to handle commodities commercially.

Experimental Procedure

This section proposes steps to be followed when developing phytosanitary treatments. Some steps may already have been considered by the time the research is done, although they may need to be substantiated for presentation to regulatory officials of importing countries.

- Develop a general outline of the proposed experiments including their objectives and how the treatment is to be carried out. Include when pests and commodities will be available, locations, types of equipment, types of treatments and project management.
- Carry out testing as required to determine response differences within the host commodity and pest populations for hosts, when evidence exists that treatment doses to provide quarantine security might differ among cultivars. For example, for hot-water immersion treatments of mango, the size and

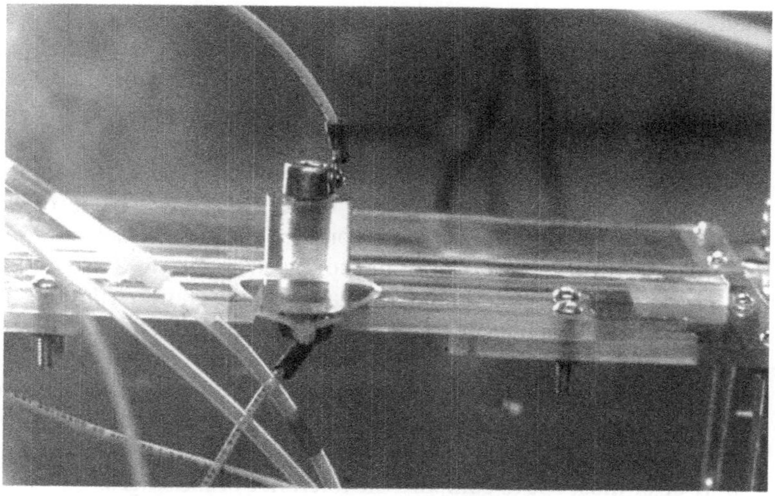

Fig. 6.3. Initial set up to test effect of pulsed electric field on fruit fly larvae and eggs that demonstrates an extreme degree of diminution that initial treatment testing may take (Hallman and Zhang, 1997). A small number of test insects are placed in a 0.8 cm^3 polycarbonate chamber. Scaling this treatment up to commercial size would involve the ability to treat pests in whole fruits and possibly other scale-up concerns unknown at this time.

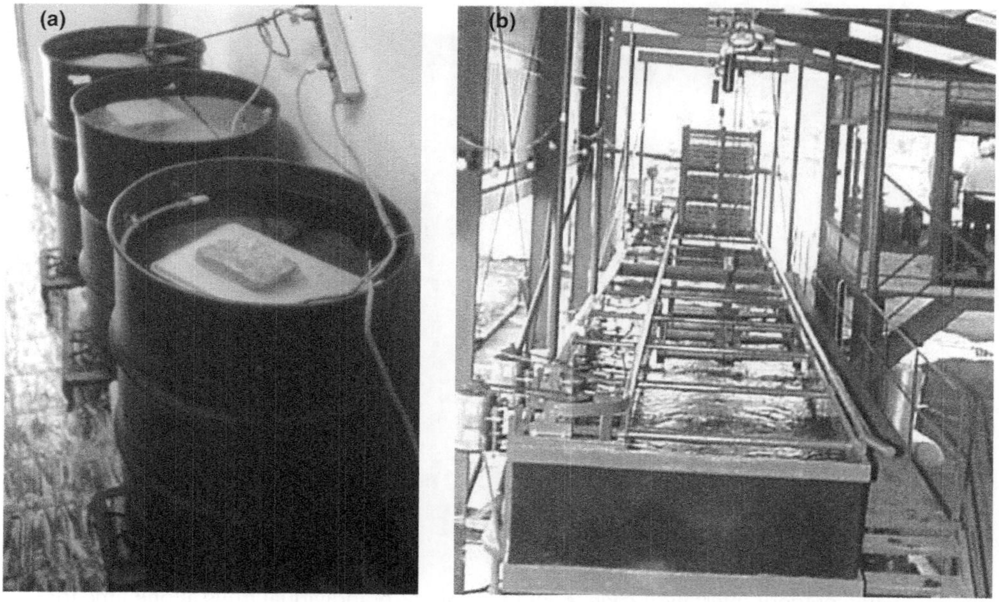

Fig. 6.4. (a) Drums used in early hot water immersion of mangoes research in Haiti to control tephritid fruit fly immatures. (b) Modern commercial hot-water immersion treatment facility for mangoes. Scaling up of the hot-water immersion treatment did not present great difficulties as the treatment follows largely the same physical heating principles between the research and commercial size facilities (see Chapter 8).

shape of different cultivars affects treatment time (APHIS, 2007). Testing for pests is necessary when conditions such as diapause exist that are possibly seasonal, causing some populations to respond differently, depending on the population from which they come. For example, plum curculio, *Conotrachelus nenuphar*, exists in two strains, one that diapauses and one that does not, with differences in susceptibility to irradiation between them (Hallman, 2003).

- Decide the type of bioassay to be used. This may be one of four types:
(i) Collection of naturally infested commodity from the field and sampling to determine the infestation levels of the stage(s) present.
(ii) Laboratory treatment of pest organisms in isolation from the commodity (i.e. in vitro). This type of test is unlikely to be acceptable alone to support a quarantine treatment proposal unless it is clearly demonstrated by preliminary testing that the result is not significantly different from treatment of the pest in the commodity, as was shown for irradiation of oriental fruit moth, *G. molesta*, in apples versus diet (Hallman, 2004a). In vitro testing may be of most use to compare the tolerance of life stages.
(iii) Laboratory infestation by permitting natural infestation of the commodity by pest organisms held in laboratory colonies or cultures, then holding the commodity until the pest develops to the life stage desired for treatment (Fig. 6.5). This is a preferable technique, because it closely mimics

Fig. 6.5. Large outdoor infestation cage to infest hundreds of individual fruits at one time for large-scale confirmatory testing. Grapefruits ready to be removed are shown. The fruits are inserted into doors on the other side of the cage and roll down inclined ramps where they are exposed to tens of thousands of fruit flies for periods of time of up to several days. This type of facility can only be used in areas where the pest of concern is not being regulated as it is far from 'insect-proof'.

the natural condition, but if infestation levels are excessive it may result in rapid decomposition of the commodity making some treatments unreliable. For example, Shellie and Mangan (2002) found that this type of infestation caused mangoes to lose 30% of their weight by the time West Indian fruit fly, *Anastrepha obliqua*, reached the third instar.

(iv) Laboratory infestation by transplanting or inoculating pest organisms of an appropriate developmental stage at an appropriate site on or in the host commodity. The pest organisms may be of field or laboratory origin. Examples include the placement of diet-reared fruit fly eggs and first instars on the inside of a plug bored into papaya and third instars in the fruit cavity for heated-air treatment experiments (Armstrong *et al.*, 1989), or injection of eggs (De Lima *et al.*, 2007) or hand placement under a flap of peel (Unahawutti *et al.*, 1992).

- Negotiate agreement with the importing country on the level of treatment efficacy (quarantine security) and the associated statistical assurance to be achieved. The following levels of quarantine security have all been used: 99.5, 99.99 and 99.9968% ('probit 9') efficacy, equating to no survivors from a nominal 600, 30,000 or 100,000 treated at a statistical assurance or CL of 95%. Required efficacy and statistical assurance jointly determine the total numbers of pests required to be treated in an experiment. What needs to be understood, especially by the national plant protection organization (NPPO) of an importing country, is what the level of efficacy means regarding risk of establishment of a new invasive pest species. The highest risk life stage of the target pest likely to be present at the time of treatment must also be agreed. Selection of the experimental model and the criterion used to determine the most tolerant stage should be made in consultation with the importing country.

The research and development of single treatments or multiple component treatments applied concurrently may be divided into two discrete phases: (i) preliminary dose-response efficacy tests; and (ii) confirmatory tests that demonstrate that a treatment is efficacious when applied to a large number (many thousands) of individuals. An appropriate dose range coupled with precision in the measurement of preliminary test doses is essential to preclude the possibility of a treatment more severe than necessary, because the maximum dose measured during the confirmatory stage of the research should normally become the minimum dose for commercial application. It is incumbent on researchers to present to the industry a treatment which is the simplest, cheapest, least injurious to the commodity and least intrusive possible while at the same time achieving the minimum level of efficacy demanded by the NPPO of the importing country.

Preliminary tests are conducted at different doses on pest stages and the results are analysed to identify the most tolerant stage. The dose may be simple as in irradiation or complex as in a combination modified atmosphere/heat treatment. Preliminary dose-range tests may be necessary to enable selection of a narrow set of doses for a subsequent experiment. This would identify the appropriate stage for large-scale testing to confirm the dose necessary to attain quarantine security. Historically, five equally spaced doses about the LD_{50} (lethal dose of a treatment to achieve 50% mortality) are usual practice. This design gives the best estimate of LD_{50} but might not be very reliable because the result sought is close to unity (complete control). For this reason, better estimates will be achieved if test doses are concentrated at the higher mortalities. Simulation studies by Robertson *et al.* (1984) and Kopittke *et al.* (2004) found that the minimum total sample size across all doses to achieve significant regressions in > 99% of simulations for a given design was 120. Asymmetric designs with a sample size of 120 gave more precise estimates of LD_{90} and > LD_{99}, although if a total sample size of ≥ 480 was used all estimates were acceptable irrespective of the design. Calculation of relative tolerance of stages across all doses is valid only if response lines do not fail the test for parallelism, otherwise comparisons are valid only at equivalent points (e.g. LD_{99}, $LD_{99.9}$). The numbers of test insects used in untreated controls must be adequate for the analysis used.

Large-scale confirmatory tests on specified numbers are usually required to obtain assurance that the treatment efficacy will satisfy quarantine security. The efficacy achieved can be determined if the number treated is known or estimated from a parallel sample. It is possible to determine the numbers which must be treated with or without survivors to achieve a required efficacy using the US Department of Agriculture software CQT-STATS (Liquido *et al.*, 1997) or the Australian AusVet software FREECALC version 2 (Cameron, 2002). Sometimes, additional commercial scale trials will be required before final acceptance of a treatment. Importing countries have differing requirements; close liaison at the planning stage is needed. Special problems may arise as discussed earlier if the pest cannot be reared in a laboratory, especially where it is not present in high densities, has disjunct distribution or is available only periodically.

RSPM No. 1 of the Asia and Pacific Plant Protection Commission (APPPC), a Regional Standard for the development of heat disinfestation treatments for fruit fly host commodities advises preliminary testing to determine the most susceptible stage using replicates of 100 insects followed by 'small scale' testing with 3000–5000 insects to determine a dose or schedule to be confirmed in a 'large scale' trial using 30,000 or more test insects. This would only satisfy requirements to demonstrate an efficacy of 99.99% at the 95% CL and would involve more research effort than the procedures outlined above have shown to be adequate. Tests against 93,600 are required to demonstrate 'probit 9' (99.9968%) at the 95% CL.

Target stage of pest

The behaviour and biology of the pest organism needs to be considered when determining the highest risk (usually the most tolerant) life stage. For example, with heat or penetrant pesticide application, where two life stages of a pest species found in a commodity are equally tolerant of a candidate treatment, the stage normally located deeper in the commodity should be selected as the target stage because treatment effects tend to be attenuated. If one of these stages is normally predominant it should be selected in preference to one rarely present. Alternatively, it can be more practicable in some situations to do large-scale tests on each relevant stage.

Dose-response regression models

Regression analyses of dose-response data may be used to determine relative susceptibility of stages. Early experimental models were based around the probit models with a log transformation of the dose (Baker, 1939; Finney, 1971). Robertson *et al.* (1984, 1994) reviewed the options available for statistical models and indicated alternative models that should be considered. Computational packages such as GENSTAT® (Anon., 2007a) or SAS (Anon., 2007b) contain appropriate routines.

Heather *et al.* (2002) compared a number of GENSTAT® dose-response models for the estimation of the heating time required to disinfest tomatoes in a hot air disinfestation system, including probit, logit and complementary log-log scales to express the probability of mortality P as a linear model. The model which best fitted the observed data was chosen and used to select the treatment time of 120 min at 44°C core temperature to achieve a minimum efficacy of 99.99%. Selection of the complementary log-log scale model in this instance was based on comparison of residual deviance and examination of the fitted curve in the upper dose range (Fig. 6.6).

The experimental design should utilize a sample unit size and number appropriate to the statistical model chosen to determine the most tolerant stage.

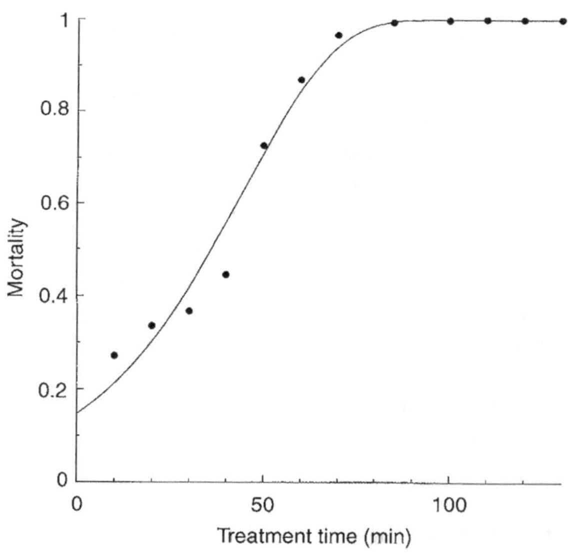

Fig. 6.6. Heat mortality response of eggs of *Bactrocera tryoni* with time at 44°C in tomatoes:

$\ln[-\ln(1-p)] = -1.841\ (0.154) + 0.04096\ (0.00305)\ \text{Time}$

where p = corrected proportion of adult mortality.

The figures in brackets after the regression coefficients in the equation are the standard errors of the coefficients (Source: prepared by R. Kopittke; reproduced with permission of the *Australian Journal of Experimental Agriculture*).

Pest organisms may be required to be treated either on the surface of the commodity or in the commodity to determine the most tolerant life stage and for prediction of the treatment dose which will be required, depending on the requirements of the importing country.

The minimum total sample size of 120 pests, but preferably 240 or more per trial can be expected to cope with biological heterogeneity. For each stage in each cultivar to be tested, three replicates with five doses, one low and the others high, should give adequate reliability of analysis when located about the LD value to be determined (Robertson *et al.*, 1984). With regression analysis, a full comparison of stages can be done only if response lines are parallel and for this a separate test of parallelism is necessary; otherwise comparisons are only valid at equivalent points, which should be in the upper range of efficacy (e.g. $LD_{99.9}$ or greater). It is not unusual to find that regression lines cross, so that a stage which is less tolerant than another at the LD_{50} can be more tolerant at the LD_{99} or higher. Of course, the most important part of the curve is at the highest levels of efficacy. If survival is the criterion for response to a treatment and the exact number of organisms actually treated is unknown, an appropriate form of analysis needs to be used such as Wadley's method (Wadley, 1949; Finney, 1971) or an equivalent (Priesler and Robertson, 1992).

Analysis of variance model

An 'analysis of variance' of responses of relevant developmental stages of the pest to a single dose can be used to determine the most tolerant stage. The dose should be one that gives few (but not zero) survivors because it is the upper level of control that is most relevant to phytosanitary security. If no survivors occur in any stage, of course, it is not possible to determine the most tolerant stage. Tests are done using at least 200 individuals of each stage in each replicate for comparison. This model is very useful for natural infestation tests because of characteristically high variance caused by uneven pest numbers in or on each sample host unit (piece of fruit or flower stem) and the difference in survival due to unit-to-unit variation. It is analogous to point-wise comparisons of LD or lethal time (LT) values necessary when regression lines for stages or cultivars are not parallel. Selection of the point of discrimination (i.e. dose) is critical as responses of stages may change relative to each other in instances where regression lines would cross at high doses.

Large-scale Confirmatory Testing

Because efficacy levels demanded of phytosanitary treatments are so close to 100%, statistically derived estimations of treatment doses are not reliable and testing of a proposed treatment on large numbers of quarantine organisms is usually a prerequisite to getting a treatment schedule approved.

Statistical considerations

It is frequently not possible logistically, nor wise biologically, to perform large-scale tests on 30,000 or more organisms in a single test and a cumulative approach must be used. Some importing countries require the large-scale test to be done as a number of 'replicates' of defined minimum size. The lowest dose at which there were no survivors in the preliminary testing can be used as the dose for the first replicate. As long as no survivors are found, subsequent testing can be done at that dose. If survivors are found the dose would be increased for subsequent tests. This represents an iterative approach and helps to ensure that the treatment is as low as possible to meet the quarantine security required. It is especially important to avoid excessive treatment on commodities susceptible to treatment injury. The dose to be used commercially should be the least dose that will achieve the required quarantine security level. If a higher dose than the effective minimum is considered necessary by a regulatory authority as a safety margin then the security level is, by implication, too low and should be renegotiated.

The number of test organisms to be used in a large-scale confirmatory test depends on the efficacy required of the treatment; which may be expressed as percentage efficacy with an associated CL typically set at 95%. Examples are given in Table 6.2. The required numbers are derived from established statistical theory discussed earlier in Chapter 5 (Couey and Chew, 1986) according to the equations for tests where there are no survivors:

$$C = 1 - (1 - p_u)^n \tag{1}$$

$$n = [\log(1 - C)]/\log(1 - p_u) \tag{2}$$

$$p_u = 1 - (1 - C)^{1/n} \tag{3}$$

where C is the confidence level (usually set at 0.95), p_u the upper possible value of the survival proportion (effectively the treatment efficacy) and n the number of test organisms. This enables calculation of the number of test insects required for a given treatment efficacy and confidence level, the efficacy achieved for a given confidence level and number of test insects, as well as the confidence level for a given treatment efficacy given the number of test insects used and an upper limit of possible survivors.

Table 6.2. Minimum number of organisms that must be tested with no survivors to achieve percentage efficacy at the 95% CL (Source: Couey and Chew, 1986; Liquido *et al.*, 1997; Cameron, 2002).

Percentage efficacy	Minimum number to test[a]
99.5	598
99.9	2,995
99.99	29,956
99.9968	93,615

[a] Numbers may differ slightly according to the calculation used and are frequently rounded up to more general figures, e.g. 100,000 for 93,615.

Where the importing country so requires, the total may need to be made up of above-mentioned minimum sized replicates, such as 7500 or 10,000 depending on the overall total and the number of replicates required. If there are survivors from larger numbers tested, the efficacy requirement may still be met with statistical validity. Some importing countries, however, may not accept a treatment from which there are survivors regardless of the number tested. Nevertheless, formulae are available for the calculation of numbers required to be treated or the efficacy achieved where there are one or more survivors (Couey and Chew, 1986; Liquido *et al.*, 1997; Cameron, 2002). Numbers required to be treated will vary slightly according to the method of calculation. For example, if one survivor is found and treatment efficacy is set at 99.9968% (C = 0.95) a minimum of 148,500 organisms must now be tested with no additional survivors instead of the 93,615 that would have been required had no survivor been found.

There is always a risk that further survivors will be found before large-scale testing is concluded which will continue to push the minimum number treated higher until the testing may become prohibitively difficult to accomplish. Therefore, it is recommended to restart large-scale confirmatory testing at a higher dose if a survivor is found that cannot be attributed to re-infestation or some other error of methodology.

The total number of individuals required to be treated without survivors in large-scale (confirmatory) trials is thus determined by the level of security required by the importing country. For example, for fruit flies the USA has historically required a level of security equal to a total treatment mortality of 'probit 9' ($LD_{99.9968}$) with 95% confidence (Baker, 1939). This requires the treatment of at least 93,600 target organisms with no survivors. Lower levels of security may be adequate for other pests in the USA. For many countries, 99.99% control is the phytosanitary quarantine requirement for all serious pests. For pests of lesser risk, some countries such as Australia and New Zealand accepted an efficacy of 99.5% (Anon., 1993) based on an inspection of 450–1250 sample units, permitting a nil tolerance from 600 samples and less than two from 950 or three from 1250.

When regression analysis is done, mortality should be adjusted by the proportion that died (or otherwise failed to achieve normal success) in the controls, commonly known in entomological circles as Abbott's formula (Abbott, 1925), before estimating levels of efficacy:

$$Y_a = 100\% - [(X - Y)/X](100\%) \tag{4}$$

where Y_a is the adjusted percentage surviving in the treated cohort, X is the percentage surviving in the control and Y is the percentage surviving in the treated cohort. For example, if control and treated survival were 80 and 20%, respectively, adjusted treated survival would be 25%.

If the level of mortality in the control is low (e.g. < 5%) adjustments may make no noticeable difference, and if they are high, depending on the organism, the test may need to be discarded. It is important to reiterate that the control must respond within the range considered to be normal for the pest species. For example, with tephritid fruit flies long reared in colonies on artificial diet it may be reasonable to expect that ≥ 80% of third instars would emerge as adults (Bustos *et al.*, 2004). Therefore, irradiation studies using prevention of adult emergence as the end point should expect that prevention of adult emergence in the control should be < 20%. If it is significantly greater than 20%, it might be necessary to study the methodology to determine the cause of the high failure rate or determine if the rate could actually be higher and still be considered normal.

Where the commodity is to be artificially infested with a known number of target organisms the number of individuals actually treated may be determined through the recovery of living and dead individuals, handling mortality can be high. If numbers of target organisms treated are unknown, as for both natural and cage-infested commodities, the number of individuals treated can be estimated using the number of individuals recovered from a control sample of infested but untreated commodity as a parallel sample (Wadley, 1949). Care must be taken in developing mortality assessment criteria. For example, normal adult fruit flies sometimes emerge from deformed puparia (Thomas and Mangan, 1995). Therefore, assuming that all deformed puparia are dead may provide inaccurate mortality data. Where the number treated is unknown it may be preferable to use data based on survivors instead of estimated mortality (Wadley, 1949; Preisler and Robertson, 1992; Harte *et al.*, 1995).

Other considerations

Ambient sensory and recording equipment used in efficacy tests must meet certified standards for sensitivity and methods of operation and be agreed upon before tests are initiated. Irradiation dosimetry must be done to a standard such as that of ASTM International (2007) for irradiation. Because treatments such as irradiation may be non-uniform across individual commodity units during treatment, the acceptable minimum treatment dose for commercial use should be set at the maximum level of the dose range in the confirmatory testing and not the mean or the target dose.

It may be required also to test a proposed treatment under actual or simulated commercial conditions. The design of this type of trial can be expected to differ for country, pest and commodity, so it will need to be negotiated with the NPPO of the importing country. Some of these trials will involve only dosimetry, for example, fumigation. Others may require that infested commodity be included and assessed for mortality.

Sequential combination treatments

Combination quarantine treatments involving sequential applications of two or more postharvest treatments, achieve quarantine security only when all treatments are applied (Mangan and Sharp, 1994). Because they are more complicated to apply than single treatments, combination treatments should be explored only when there are compelling reasons, such as no single tolerated treatment will provide acceptable quarantine security alone. For example a commodity may be susceptible to injury from a treatment that would be effective alone at a higher dose or if different stages of a pest's life cycle are the most susceptible to different treatments. Examples of sequential combination treatments include methyl bromide fumigation and cold storage (APHIS, 2007), irradiation and cold storage (Castro *et al.*, 2004), and compression and phosphine fumigation (Yokoyama and Miller, 2003). Combination treatments may have additive or synergistic effects on pest mortality. Some prospective combination treatments may actually be no better than one of the components alone, such as cold and modified atmospheres (Chapter 11), and some combination treatments could be potentially antagonistic, such as irradiation and modified atmospheres (Chapter 9). Therefore, the fact that two or more treatments are combined does not ensure a more efficacious treatment.

Statistical confidence limits would be based on pest survival for the combined treatments proposed. Injury responses of the commodity should be reported for all treatments used to provide quarantine security. Where chemical treatments are combined with physical treatments residue analysis data will be necessary.

Mangan and Sharp (1994) briefly discuss two separate approaches to developing combination treatments. One is simply an iterative, rather serendipitous approach to evaluate the effect of each treatment separately and then attempt to combine the effects of both at the most optimal doses of each for efficacy and commodity quality. The other uses a statistical approach to identify dose levels of each treatment which when used together yield synergism in efficacy while maintaining quality. Identification of synergism is the preferred method for developing a combination treatment because it may require less expenditure of treatment resources to achieve quarantine security. The development of modified atmosphere/heat combination treatments has probably been the most successful employment of synergism accomplished in phytosanitation (Neven and Rehfield-Ray, 2006; Neven *et al.*, 2006b), although these treatments have not been used commercially yet (Chapter 11).

The following sequence is one way of developing a multi-component quarantine treatment:

- Determine the most tolerant developmental stage of the target pest that would be encountered at the time of treatment for each of the components of the combination treatment. Stages that are found inside the commodity should be treated inside it. When the most tolerant stage to each component of the treatment is the same stage, further tests including large-scale confirmatory testing are done with the combined treatment as a whole. If the most tolerant stages are different for different treatment components, a separate series of preliminary tests may be required to determine the treatment doses for each component of the confirmatory test. If previous research indicates that the first treatment in the sequence achieves the necessary quarantine security against certain developmental stages, only the surviving stages need to be tested in subsequent treatments.
- Determine pest mortality for a range of treatment doses for each component to estimate the maximum number of survivors at the highest doses which can be tolerated by the commodity. Based on these results a multiple treatment schedule may be proposed which will provide quarantine security. It may be the highest minimum of the potential commercial max/min ranges.

- Confirm that the proposed multiple treatment schedule provides quarantine security against the most resistant developmental stage of the target pest under actual or simulated commercial conditions. In cases involving treatments that target different developmental stages of the pest, separate confirmatory tests on each stage may be required. The sequence of treatments used in the confirmatory tests will be the sequence proposed for commercial use.

Postharvest Commodity Quality

It is essential that postharvest quality studies be done on commodities in conjunction with pest disinfestation testing to ensure that treatments do not result in unacceptable levels of injury to the commodity. Commodity tolerance is of primary importance for fresh commodities and propagative materials as they are alive and actively metabolizing as are the pests infesting them. Several reviews discuss commodity quality related to phytosanitary treatments (McDonald and Miller, 1994; Paull and Armstrong, 1994; Lurie, 1998; Molins, 2001; Tang *et al.*, 2007).

Commodity tolerance is not of direct concern to regulatory agencies although they make an effort to provide phytosanitary treatments that are considered to yield commodities of acceptable quality. Regulatory agencies deny responsibility for damage caused to commodities by treatments that they accept. It is up to those with economic interests in the commodity being treated to be comfortable with the quality of commodities treated on a commercial scale, which might differ from those treated on a small, experimental scale. Pilot commercial trials to evaluate commodity quality under commercial conditions of export and storage should be conducted.

Three criteria that are most important in evaluating treated edibles are gross appearance, organoleptic qualities and shelf life. Shelf life is not noted for most commodity quality studies, and organoleptic evaluation is often lacking. Quality tests may be done without sophisticated scientific equipment by treating lots of at least 20 individual commodities with treatment conditions approximating commercial conditions and storage temperatures and times. Studies should be replicated across the bulk of commercial commodity variables including cultivar, growing area, season, size or any factor that may affect quality. Properly constituted panels doing evaluations (Amerine *et al.*, 1965) should not know which lots were treated and which were controls and these evaluators are ideally composed of individuals whose 'job' includes evaluating the quality of commodities, such as produce brokers, importers, retail managers and informed consumers. They would judge the commodity based on criteria that they would use in deciding to buy or consume it: appearance, both exterior and interior, firmness, texture, both in the hand and in the mouth, smell and flavour. Detectable differences may be inconsequential. Shelf life would be evaluated by holding a part of the lot at typical storage conditions for that commodity, be it in a storage facility, retail store or home, until the commodity is no longer useful.

Non-edible commodities would be evaluated similarly, except for taste, of course. Cut flowers may put emphasis on appearance, shelf life and perhaps fragrance. Experimental measurements (e.g. Brix, pH, colour, penetration, citric acid content, electrolyte leakage) are not necessary unless certain conditions that require such measurements are being investigated. For example a consumer group wishes to know if a treatment affects vitamin content or a researcher is investigating why a fruit loses considerable shelf life after treatment.

Even if results of quality trials have been reported by researchers, it would behove industry to do some type of quality trials themselves to be sure that they accept any effect the treatment has on the commodity. Industry's criteria for acceptance may be harsher or milder than the researchers'. Doses should be extended to a level which will identify the threshold for commodity injury, which should be the upper limit permitted.

The following are specific parameters which could be used to assess possible injurious effects of treatments on commodities:

- Colour development or changes.
- Firmness of fruit or flower tissue.
- pH of fruit juice.
- Brix (total soluble solids) of fruit as a measure of maturity.
- Ascorbic acid content of fruit.
- Ethylene production patterns with time.
- Carbon dioxide production patterns with time.
- Visible injuries.
- Chlorophyll fluorescence.
- Citric acid content of fruit.
- Eating quality of produce.
- Appearance of cut flowers.
- Shelf life.
- Decomposition.

Data on these parameters should be collected using sound experimental designs and should be analysed statistically. It should be done from a representative range of production areas and repeated over more than one season, ideally three, comparing with untreated commodity otherwise held identically.

Submission of Proposals

The SPS Agreement (WTO, 2007a) formulated under the auspices of the WTO provides overall guidance for negotiation of market entry of commodities subject to phytosanitary quarantine constraints. Likewise, the International Plant Protection Convention (IPPC) is the phytosanitary standard setting body identified in the SPS Agreement. The IPPC (2007) produces International Standards for Phytosanitary Measures (ISPM) that provide specific guidance on the formulation of phytosanitary measures, including phytosanitary treatments and systems approaches. The format of a proposal for assessment of a dis-infestation treatment depends on the requirements of the importing country's NPPO and should be prepared in consultation with that country. Especially for irradiation, but usually for any method, full records including timings, dosimetry results to an approved standard and any other relevant parameters should support the proposal. If possible, skilled technical advocates should present a submission to ensure that there are no misunderstandings.

The following tables of data should be included with a submission:

- Development times for stages of the pest in culture medium used for in vitro tests, if any, range of times and/or mean or modal times ± standard error of the mean (SE).
- Development times for stages of the pest in each variety of the fruit or flower proposed for export, range of times and/or mean or modal times ± SE.
- Treatment mortality or survivors of each stage of each pest species at doses resulting in near 100% control.
- Dose-response analyses on each stage of each pest species that may be present in the exported commodity; these should include appropriately transformed data with 95% CL (or fiducial limits), parallelism-test results if necessary with comparisons of lines. Results should meet the assumptions intrinsic to the model. Analysis of variance at one or more doses comparing responses of stages could be given.
- The size range of commodities, such as dimension or weight.
- Results of any tests to predict the dose needed for the required quarantine security (e.g. 99.5%, 99.99% or 99.9968%) on the most tolerant stage in the most susceptible cultivar.
- Results of large-scale dose confirmatory trials.
- For specific treatments, details of the treatment facilities, load, packaging, dose measurements and calibrations, and performance monitoring data during the treatment should be included.

Conclusions

Increasing formalization of specifications for phytosanitary requirements worldwide will require increasingly precise and complex experimentation and submissions to satisfy these requirements (Follett and Neven, 2006). This will be of positive value to regulatory organizations that will be able to place greater trust in presentations and facilitate their interpretation. Although this will place high demands on research staff responsible for the work in exporting countries it will enable them to work within a more predictable set of guidelines. The IPPC programmes including its ISPM will provide excellent guidance to countries initiating trade in commodities subject to phytosanitary constraints for the first time. However, possibly its greatest contributions overall will be through harmonization of requirements and fostering of the principle of equivalence wherever it is applicable.

7 Disinfestation with Cold

Holding commodities at temperatures near the freezing point of water was one of the first major methods for disinfestation of quarantine pests. It was used during the first incursion of Mediterranean fruit fly, *Ceratitis capitata*, in Florida in 1929 and has been used against a wide variety of pests on numerous commodities ever since.

Ironically, the effect of cold on insects in exported fruit was first noted because it was found to prolong the life of pest eggs and larvae in fruit shipped long distances under cold storage (Gould, 1994). The widespread use of cold to ship fresh commodities long distances is likely to have contributed to the spread of invasive species since its implementation in the late 19th century. The first studies of the use of cold to kill insects in fresh commodities were not as a phytosanitary treatment to overcome quarantine barriers but simply to stop the spread of invasive species via commodities shipped under cold storage. Only later was cold studied and used as a phytosanitary treatment per se.

An advantage of cold treatment is its tolerance by a wide variety of fruits, including many tropical and subtropical ones. Some fruits, such as apples, are stored for months at temperatures that are lethal to most quarantine pests, so that disinfestation can be accomplished during postharvest storage. Furthermore, cold treatment can be applied to fruits after packing and 'in transit' during lengthy transport by sea.

A cold treatment system is usually relatively inexpensive to install compared with other treatments, especially if cold storage facilities already exist and only need to be certified as able to produce the temperatures needed uniformly and without unacceptable fluctuations. Some upgrading of equipment might be required. Cold treatment is free of chemical residues and is accepted by organic growers, a group whose market share, although a fraction of the total, steadily increases.

The chief disadvantage of non-freezing cold disinfestation is the long treatment time. No other treatment, with the exception of some modified atmosphere grain treatments that are not done against quarantine pests in any case, requires such long time periods (up to several weeks) to be effective. Lengthy treatment time exposes cold to greater risk of interruption caused by equipment malfunction or power outages. If a cold treatment is interrupted and the temperature rises by as little as a degree for even a relatively short period of time the treatment may be disallowed and re-treatment required or the shipment rejected. Cold commodity disinfested in transit can be prone to local 'hot spots' which need to be identified by monitoring if the possible occurrence of pest survivors is to be avoided.

Thermometry

The Celsius temperature scale is the accepted international standard for biological research and technology. However, for quarantine regulatory purposes the USA uses the Fahrenheit temperature scale (APHIS, 2007). Conversions of temperatures back and forth between the two scales often result in values of up to two decimal places being published. This degree of apparent precision resulting from the conversion of readings should be avoided as it gives the false impression that the treatments were done or are being regulated to a more precise degree than the research supports. Also, despite advances in electronic sensors and control circuitry, few commercial capacity quarantine cold treatment facilities would be capable of temperature calibration intervals smaller than 0.1°C and a realistic variance range for a treatment facility would be ± 0.5°C. The Animal and Plant Health Inspection Service of USDA (APHIS) allows a variance of 0.3°C approximating to 0.5°F above the required temperature (APHIS, 2007).

If the treatment temperature is given as a value with a variance allowable in recorded readings the commodity core temperature adjustment should be set to achieve the treatment temperature as a mean. Alternatively, if a maximum temperature is given, allowance must be made for the variance by setting a temperature at or below the maximum minus the variance of the equipment. In addition to equipment variance, refrigeration equipment almost invariably requires a defrosting cycle about once in every 24 h during which the cooling air entering the chamber will rise above the set value. This should have negligible effect on the commodity bulk temperature and is accepted by cold treatment protocols which may accept chamber temperature measurements taken within water-filled containers no larger than a litre inside the treatment chambers that buffer short-term temperature changes.

Temperature is typically controlled and monitored from data obtained from resistance temperature detector probes, thermocouples or other sensors of similar precision. These would be located according to temperature gradients in the chamber mapped during a prior calibration/certification procedure. Probes monitoring fruit pulp temperature should be placed in the largest fruit if there is a significant variation in size within the load.

Physiology of Cold Tolerance in Pests

Freezing temperatures are a special challenge to insects. They have evolved a variety of physiological adaptations to prevent damaging ice from forming in the body (Denlinger and Lee, 1998). Although mechanical damage by ice crystals may be a major mortality factor, the excessive concentration of extracellular fluids driven out of the growing ice lattice, resulting in osmotic removal of yet unfrozen cellular water and cellular dehydration may be the primary causes of physiological stress. Freezing may result in shrinking of cells to the point where the membrane is damaged beyond repair once temperatures return to ambient.

Most phytosanitary cold treatment does not involve freezing. Mortality caused by cold without the production of ice is less well understood than that caused by freezing. Enzymatic activity decreases as the temperature lowers in poikilotherms and changes in the tertiary structure of proteins and subsequent disassembly of polypeptide units result in protein denaturation. Damage to the plasma membrane may occur because chilling induces cell fluid to the gel phase resulting in alterations in membrane permeability, reduction in the activity of membrane-bound enzymes, and permanent separation of membrane proteins and lipids (Denlinger and Lee, 1998). This is more likely to happen in tropical species that are not normally exposed to cold and do not have extensive cold adaptations.

Most Cold-Tolerant Stage

As with all phytosanitary treatments, the most tolerant pest stage present in the shipped commodity should be identified and the treatments should be proven to be adequately efficacious against that stage. The most tolerant stage should be determined using a common objective. For tephritid fruit flies, in particular *C. capitata*, results of studies to determine the most cold-tolerant stage have produced conflicting results. This may be because *C. capitata*, arguably the most studied pest for phytosanitary solutions over the longest period of time, was subjected to a wide variety of research methods and qualities.

Back and Pemberton (1916) did the first extensive cold work on a fruit fly (*C. capitata*) and found variable results depending on the temperature (Table 7.1). First and third instars were generally the most tolerant at temperatures in the range of those mostly used for phytosanitary treatment of fruit fly hosts (0–2.2°C) while third instars were more tolerant above that range. However, numbers of insects used in this study for each temperature/time combination varied from < 100 to < 1500, numbers that are insufficient to confirm phytosanitary treatments at a satisfactory efficacy and level of statistical confidence (Couey and Chew, 1986; Liquido *et al.*, 1997). The values in Table 7.1 would probably be raised by at least 20% if large-scale confirmatory testing were done.

An interesting result of Back and Pemberton (1916) is that from 0°C up through the 2.2–4.4°C range changes in temperature seem to have no effect on mortality. It is as if a threshold is reached around 4.4°C where further lowering of

Table 7.1. Mortality of Mediterranean fruit fly exposed to cold (Source: Back and Pemberton, 1916).

Temperature (°C)	Days to achieve 100% mortality per stage			
	Egg	1st instar	2nd instar	3rd instar
0	10–11	9	10	12
0–0.6	9–10	11–12	9	10
0.6–1.19	17	11	12	12
1.1–2.2	⩽ 11	⩽ 10	⩽ 8	13
2.2	12	10	8	10
2.2–4.4	⩽ 9	⩽ 11	8	10–11
3.3–4.4	16	17–20	21–25	> 28
4.4–7.2	21	–	30–31	46

the temperature has little effect on mortality until freezing is achieved. If that were true it would behove industry to use the upper end of this temperature range to save resources and reduce chilling injury to the commodity. The anomaly of 17 days to achieve 100% mortality of first instars at 0.6–1.1°C, while lower times achieve the same level of mortality above that temperature, may be due to statistical variation and the relatively small sample sizes for measuring mortality near 100% and further supports the contention that days to achieve quarantine security against *C. capitata* are higher than the values presented in Table 7.1.

Powell (2003) reported that the data from Back and Pemberton (1916) 'informed development' of the cold treatment schedules used by USDA-APHIS prior to 2002 (10, 11, 12, 14 and 16 days, respectively, at 0, 0.6, 1.1, 1.6 and 2.2°C), although it must be acknowledged that other papers available at the time, perhaps even some unpublished, may have influenced the schedules. In an unpublished analysis of the Spanish clementine case, W. Gould (personal communication) supposed that Nel (1936) was an important source. The data of Back and Pemberton (1916) show that the time at 0, 0.6 and 1.1°C should have been at least 12, 12 and 17 days, respectively, and that any treatment at 1.6°C is not supported by the data, as Back and Pemberton (1916) reported no studies at that temperature. Nel (1936) reports data at 1.6°C; but concludes that 16 days are needed. He further recommended 9 days at 0.6°C and 12 days at 1.1°C. In any case, the data of Nel (1936) are insufficiently robust to support phytosanitary treatments by themselves. A small number (4–34) of unspecified fruit were used for each data point and numbers of insects treated is not given. Furthermore no control was used.

Powell (2003) analysed the data of Back and Pemberton (1916) and suggested that in the temperature range of 0–2.2°C, duration of the treatment may be more important than temperature. Powell (2003) notes that given the age of the data, when cold storage capabilities for long durations were not as precise and reliable as they are today, and incomplete methodology reported in Back and Pemberton (1916), results from the analysis should be considered hypothetical and further research to test the hypotheses done, especially to verify efficacy of low-temperature short-duration treatments. Powell (2003) compares later results of cold treatment studies of *C. capitata* and finds that the data generally agree with his analysis, although he did not find any data sets as detailed as that of Back and Pemberton (1916), and further research in this area is needed. Current cold treatment schedules for *C. capitata* in the APHIS Treatment Manual (APHIS, 2007) are for 1.1, 1.6 and 2.2°C (Table 7.2). The lower temperature treatments (0 and 0.6°C) have been removed unless further research can confidently restore them.

A data set that Powell (2003) did not review is Armstrong *et al.* (1995), who found that first instar *C. capitata* might be more cold tolerant than the third instar (two of 120,896 first instars pupariated after exposure to 1.1°C for 12 days while none of 30,805 third instars did so). This observation is in agreement with Back and Pemberton (1916) at that temperature.

A data set published since Powell (2003) by De Lima *et al.* (2007) found that the second instar of *C. capitata* was consistently the most tolerant stage to 2 and 3°C when tested on five varieties of citrus fruit (Table 7.3). Host affected tolerance; the LD_{95} for second instar in 'Lisbon' lemon was 9.7 days, while it was 13.6 days in 'Valencia' orange. In the same study, Queensland fruit fly, *Bactrocera tryoni*, first instars were the most tolerant stage and *B. tryoni* was less tolerant of cold than *C. capitata*. In this study, temperature (2 or 3°C) did not seem to matter for *B. tryoni*, extending the theory of Powell (2003), that temperature does not matter as much as time between 0–2.2°C for *C. capitata*, to *B. tryoni* for temperatures at least up to 3°C. In the study of De Lima *et al.* (2007) 95% prevention of pupariation took longer at 3°C compared with 2°C for *C. capitata.*

The physiology of cold tolerance in insects is complex and much remains to be understood. Near the low temperature threshold for survival of *C. capitata* the response to time at a temperature is more sensitive than the response to temperature difference (Powell, 2003) and this can be expected to apply to other Tephritidae. This is possibly indicative of a lethal mechanism with a temperature threshold. Several mechanisms are known to be involved in high cold tolerance including accumulation of polyols, thermal hysteresis and removal or inactivation of ice-nucleating agents.

A few species of insects may sustain supercooling to −20°C or less, making deep-freezing as a disinfestation technique very difficult to use against them (Zacchariassen, 1985). For grain pests, at extended storage temperatures around

Table 7.2. APHIS cold treatment schedules for Mediterranean fruit fly before and after finding live larvae in cold-treated mandarins in the USA in 2001.

Maximum temperature (°C)	Pre-2002 treatment durations (days)	Post-2001 proposed treatment durations (days)	Present treatment durations (days)[a]
0	10	12	(Not allowed)
0.6	11	13	(Not allowed)
1.1	12	14	14
1.6	14	16	16
2.2	16	18	18

[a] Source: APHIS (2007).

Table 7.3. Estimated LD_{95} (95% confidence limits) for Mediterranean fruit fly, *Ceratitis capitata*, and Queensland fruit fly, *Bactrocera tryoni*, in five citrus fruits at two temperatures. End point is failure to produce normal-looking puparium (Source: De Lima *et al.*, 2007).

		LD_{95} (95% confidence limits) in days at			
		2°C		3°C	
Fruit	Insect stage	*C. capitata*	*B. tryoni*	*C. capitata*	*B. tryoni*
'Valencia' orange	Egg	9.8 (9.7–9.9)	4.7 (4.5–4.9)	9.2 (9.3–9.4)	4.2 (4.0–4.3)
	1st instar	12.7 (12.5–12.9)	7.9 (7.6–8.2)	12.9 (12.8–13.1)	6.6 (6.3–6.9)
	2nd instar	13.6 (13.4–13.8)	5.3 (5.1–5.5)	14.7 (14.5–14.9)	5.4 (5.2–5.6)
	3rd instar	9.8 (9.7–10.0)	6.0 (5.8–6.2)	10.9 (10.8–11.1)	5.0 (4.9–5.2)
'Navel' orange	Egg	8.9 (8.8–9.0)	5.3 (5.3–5.5)	9.8 (9.7–9.9)	7.5 (7.3–7.8)
	1st instar	12.5 (12.3–12.7)	7.2 (7.0–7.4)	13.8 (13.7–13.9)	7.9 (7.7–8.1
	2nd instar	12.6 (12.5–12.8)	5.4 (5.2–5.6)	14.0 (13.8–14.1)	6.2 (6.1–6.4)
	3rd instar	10.3 (10.2–10.5)	6.4 (6.2–6.6)	11.0 (10.9–11.2)	6.7 (6.5–6.9)
'Lisbon' lemon	Egg	8.5 (8.4–8.6)	3.4 (3.3–3.6)	9.0 (8.9–9.1)	3.3 (3.1–3.5)
	1st instar	9.7 (9.6–9.8)	3.9 (3.8–4.1)	11.1 (11.0–11.3)	3.8 (3.7–4.0)
	2nd instar	9.7 (9.6–9.9)	3.7 (3.6–3.9)	12.1 (12.0–12.3)	3.5 (3.3–3.6)
	3rd instar	8.9 (8.8–9.1)	3.3 (3.1–3.5)	10.0 (9.9–10.1)	3.0 (2.7–3.2)
'Ellendale' mandarin	Egg	9.1 (9.0–9.2)	3.6 (3.6–3.7)	9.9 (9.8–10.0)	4.3 (4.2–4.4)
	1st instar	11.3 (11.2–11.5)	7.7 (7.6–7.8)	13.1 (12.9–13.3)	7.5 (7.4–7.6)
	2nd instar	11.7 (11.6–11.9)	6.7 (6.6–6.8)	13.4 (13.2–13.6)	7.2 (7.1–7.3)
	3rd instar	10.0 (9.9–10.1)	5.2 (5.1–5.2)	11.0 (10.9–11.2)	6.8 (6.7–6.9)
'Murcott' mandarin	Egg	8.4 (8.3–8.5)	4.6 (4.5–4.7)	8.8 (8.6–8.9)	5.0 (4.9–5.1)
	1st instar	10.9 (10.8–11.0)	7.9 (7.8–8.1)	12.7 (12.6–12.9)	7.2 (7.1–7.3)
	2nd instar	11.7 (11.5–11.9)	6.3 (6.2–6.4)	13.3 (13.1–13.5)	5.3 (5.2–5.4)
	3rd instar	9.6 (9.5–9.8)	6.3 (6.2–6.4)	10.4 (10.3–10.6)	4.8 (4.7–4.9)

15°C population increase tends to be inhibited and below 8–10°C eventual disinfestation will occur. Cold storage of fruit and vegetables, for other purposes including when part of longer term modified atmosphere storage (Hallman, 1994a), will eliminate most species of pests and provide acceptable quarantine security.

Temperate insects that undergo diapause usually require longer cold treatment times than tropical and other insects that do not diapause. For example, at 2.2°C, the tropical, non-diapausing Oriental fruit fly, *Bactrocera dorsalis*, requires 10 days while the temperate, diapausing apple maggot, *Rhagoletis pomonella*, requires 42 days (FAO, 1984) to achieve quarantine security.

Cold acclimation in insects is a physiological phenomenon which can influence the efficacy of a disinfestation treatment. Development of cold acclimation during cooling down was recognized by Iwata *et al.* (1992) in a study of cold tolerance in *B. dorsalis*. Meats (1976) reported cold acclimation in *B. tryoni* but his study related to climatic adaptation temperature patterns, not disinfestation. Evans (1981) found cold acclimation expressed as a lowering of the chill coma temperatures in several stored grain pests, especially the sawtoothed grain beetle, *Oryzaephilus surinamensis* (family Silvanidae) at temperatures relevant to refrigerated aeration storage regimes. Cold acclimation does not appear to have been studied in *C. capitata*, the most important tephritid fruit fly pest species, but it could account for some of the differences in published data, which have given rise to prescribed phytosanitary treatments.

Influence of Cool-down Time

Cold disinfestation requirement is normally prescribed as a period of time after all of the commodity bulk has reached the nominated treatment temperature. In commercial operations many fresh fruits and vegetables are cooled immediately after harvest to avoid loss of quality so the mortality during cooling-down (pre-cooling) time to reach the quarantine disinfestation temperature will be relatively low due to the small time contribution, especially where a forced-air cooling system is used. However, for large bulks of commodity in cooling chambers, which rely on convection as opposed to forced-air cooling, the cool-down time can be expected to contribute significantly to the overall mortality. However, slow cool down may enable physiological acclimation by the pest to occur which could increase its tolerance to the treatment (Iwata *et al.*, 1992).

Research to demonstrate the efficacy of cold treatments differs from commercial practice in that during research fresh produce is maintained at temperatures favourable to pest survival after treatment in order to check for survivors. The commercial practice of cooling produce soon after harvest and maintaining it cool after any cold phytosanitary treatment may augment the lethal action of a cold treatment. Therefore some regulatory cold treatment requirements may over achieve in terms of quarantine security.

In experiments there will be an unknown cooling-down time contribution to mortality as not all fruit in experimental lots will be cooled at the same rate and this adds to sources of error which affect experimental precision. For example, in the disinfestation experiments of De Lima *et al.* (2007) against *C. capitata* and *B. tryoni*, selected *Citrus* spp. infested with *C. capitata* were treated together with buffer fruit in standard export cartons containing 19 kg and cooling from 26°C to 2 or 3°C took 42–46 h. Fruit infested with *B. tryoni* were treated similarly but infested in a different manner at a different laboratory facility remote to that for *C. capitata*, which could have led to some distortion in statistical comparisons. While the effect of cool-down time may be more or less comparable across experiments in a given laboratory for a given commodity and sample size it may differ between laboratories due to differences in equipment, and acclimation could subsequently be induced in some treatment schedules. Consequently, schedules that require large time differences between closely related pest species or even between differing market requirements for the same species may be attributable to differences in the way the research was conducted.

Because quarantine is primarily concerned with achievement of a given security standard, commercial facilities for the treatment of critical pests can be expected to require individual confirmatory testing and approval by regulatory authorities of the importing and exporting countries. While the schedules adopted may achieve the security required they might not necessarily be those which involve the minimum possible exposure to disinfestation cold temperatures.

Commodity Tolerance

Although temperate fruits generally tolerate cold better than tropical fruits, so do temperate insects, requiring longer treatment times to achieve quarantine security compared with tropical insects. Subtropical citrus fruit can exhibit cold injury after treatment at 0–2°C and may require curing at ambient temperature before cold disinfestation treatment followed by normal storage at 10°C if chilling injury is to be minimized (McLauchlan *et al.*, 1993). Most tropical fruits do not tolerate the cold temperatures and times required for phytosanitary treatments ($\leqslant$ 3°C for 10–40 or more days) although some, such as canistel, white sapote and carambola, do (Hallman and Chalot, 1993).

Temperatures of 10°C can be lethal to tropical fruit flies given long exposure times (several weeks) but 3.3°C is about the highest temperature suggested as a phytosanitary cold treatment in order to accomplish the treatment in a reasonable period of time, such as 3 weeks maximum for the most cold-tolerant tropical fruit flies (FAO, 1984; APHIS, 2007). For pests such as codling moth, *Cydia pomonella*, cold treatment alone at above-injury thresholds is inadequate (Moffitt, 1971) and must be combined with fumigation, again at above injurious dosage thresholds.

Although cold is one of the most widely used phytosanitary treatments it has not been researched to any significant degree for use on cut flowers and foliage even though many of these commodities tolerate the temperatures and time periods required to kill insects (Hardenburg *et al.*, 1986). Typically, cut flowers are transported as air cargo and a long cold treatment does not fit current commercial practices. One example of the successful use of cold is to disinfest strawberry planting material, 'runners', against western flower thrips, *Frankliniella occidentalis*, by Williams *et al.* (2005) with cold treatment at −2°C for 4 weeks.

Combination Cold–Methyl Bromide Fumigation Phytosanitary Treatments

Cold preceded or followed by methyl bromide fumigation is certified as a phytosanitary treatment for certain tephritid fruit flies, Lepidoptera and mites (FAO, 1984; APHIS, 2007). Because a combination treatment is generally more complicated and costly than a single treatment there should be an overriding reason for using it instead of a single treatment. One reason is for commodities that may not tolerate a single treatment at an efficacious dose. For example, methyl bromide fumigation at 32 g/m^3 for 2–3 h, followed by cold treatment at temperatures and times varying from 0.6–2.8°C for 4 days to 8.9–13.3°C for 10 days, with the shorter fumigations coupled with more severe cold treatments, is used to disinfest avocado and several other fruits from some tephritid fruit flies and grapevine moth, *Lobesia botrana* (APHIS, 2007). Avocado generally does not tolerate a single cold or methyl bromide treatment sufficient to control these pests.

Another reason for using a fumigation/cold combination treatment is to take advantage of relatively short transit times to do the cold treatment. For example a cold treatment alone against Mediterranean fruit fly at 2.2°C requires 18 days. Regarding the combination methyl bromide/cold treatment following fumigation with 32 g/m^3 methyl bromide for 2 h, the cold component in that temperature range would be for 4 days (Table 7.4). Given a transit time of about 4 days the commodity would be ready for market upon arrival. By contrast, for the single cold treatment the commodity would need to be cold-stored for an additional 2 weeks for completion of the treatment and release to market. The shorter cold treatment would also be less likely to be interrupted by equipment malfunction or power outages.

Cold Treatments for Specific Pests and Groups

Phytosanitary cold treatment schedules exist for a number of pests including tephritid fruit flies, lepidopterous borers and weevils. Research has been done for other pests.

Mediterranean fruit fly

Tropical fruit flies are the primary group of quarantine pests for which phytosanitary treatments are designed, and foremost among these is the Mediterranean fruit fly, *C. capitata*. The first dedicated cold disinfestation treatment was against this pest in Florida in 1929 (Richardson, 1952). This treatment of 1.1°C for 12 days was tolerated by citrus fruit and was intended to achieve a minimum efficacy equivalent to 99.99%. However, the USA subsequently raised the efficacy required for quarantine security to 99.9968% ('probit 9') which by and large serves as the basis for treatment schedules against fruit fly in the USA to this day. Treatment schedules at other temperatures were approved (Table 7.2).

In late November of 2001 live Mediterranean fruit fly larvae were intercepted in cold-treated clementines (a type of mandarin, *Citrus reticulata*) from Spain that

Table 7.4. Methyl bromide fumigation (32 g/m^3) followed by cold storage combination treatment against Mediterranean fruit fly (Source: APHIS, 2007).

Duration of fumigation (h)	Cold treatment temperature ranges (°C)	Cold treatment duration ranges (days)
2	0.6–8.3	4–11
2.5	1.1–13.3	4–10
3	6.1–13.3	3–6

had been purchased by consumers in Maryland and North Carolina, states that are homes to the APHIS, Plant Protection and Quarantine headquarters and the Eastern Regional Office, respectively (APHIS, 2002a). On 30 November 2001, APHIS notified the government of Spain that importation of clementines was being suspended pending investigation. On 5 December 2001 shipments were resumed after APHIS identified one ship as being the source of infested fruit and suspected that the cold treatment was not being done properly by that ship. Later on the 5 December inspectors in Louisiana found more live Mediterranean fruit fly larvae in clementines from Spain that had originated on a different ship. Further interceptions were made around that same period in California and New Jersey. Clementine imports from Spain were subsequently restricted to north-eastern states where Mediterranean fruit fly host material was absent during that time of the year.

This number of interceptions in a short time span was unprecedented in the history of the legal importation of fruit hosts of the Mediterranean fruit fly to the USA. Infestation levels in Spanish clementines that year were uncharacteristically heavy and may have overwhelmed the cold treatment or the treatment may not have been applied to provide the expected level of efficacy. It was decided that efficacy should not be placed entirely on the cold treatment but that measures to keep Mediterranean fruit fly infestations relatively low should be implemented. Fruit fly traps, bait sprays and pre- and post-treatment sampling would help determine and maintain low infestation levels. This type of phytosanitary philosophy should be applied to all treatments where the treatment becomes part of a system designed to maintain the risk of quarantine pest establishment at acceptable levels. An increase in cold treatment times (Table 7.2) and more detailed and precise recording and examination of treatment records were changes in the treatment regime designed to forestall future problems with cold treatment against Mediterranean fruit fly. Later, cold treatments at the lower temperatures were dropped because subsequent analysis of existing data deemed those treatments to have an unacceptable risk of failure (Powell, 2003). No live larvae have been found in cold-treated Spanish clementines since these changes were implemented in 2002.

It is encouraging from the standpoint of the establishment of invasive species that no infestations of Mediterranean fruit fly in the USA seem to have resulted from this apparent widespread distribution of host fruit with live larvae. This may be in part because the larvae, although alive, were moribund from the treatment. It also brings into question how rapidly the cold treatment causes mortality. This issue has been raised with heat treatments and fumigation also (see Chapters 8 and 10, respectively) and is due in part to the fact that cold treatment research is often done in a way that allows for live larvae but not some latter stage, such as puparia (Powell, 2003). Inspectors generally count any live larva as a treatment failure, except ionizing irradiation (Chapter 9).

Other tephritid fruit flies

Examples of cold treatment schedules for other tephritids besides Mediterranean fruit fly are given in Table 7.5. *Bactrocera* spp. tested have been consistently more cold susceptible than Mediterranean fruit fly and like the Mediterranean fruit fly,

Table 7.5. Cold disinfestation schedules for some quarantine pests.

Pest group and species	Temperature (°C),[a] days	Market[b]	Comments, references, commodities
Fruit flies (Diptera: Tephritidae)			
Anastrepha ludens	0.6 ± 0.3, 18		
	1.1 ± 0.3, 20		
	1.7 ± 0.3, 22	USA	APHIS (2007), Schedule T107-b
Anastrepha spp. except *A. ludens*	0 ± 0.3, 11		
	0.6 ± 0.3, 13		
	1.1 ± 0.3, 15		
	1.7 ± 0.3, 17	USA	APHIS (2007), Schedule T107-c
Bactrocera tryoni	0 ± 0.3, 13		
	0.6 ± 0.3, 14		
	1.1 ± 0.3, 18		
	1.7 ± 0.3, 20		
	2.2 ± 0.3, 22	USA	APHIS (2007), Schedule T107-d
	2 ± 0.5, 16		
	3 ± 0.5, 18	Japan	De Lima *et al.* (2007), lemon
	2 ± 0.5, 18		
	3 ± 0.5, 20	Japan	De Lima *et al.* (2007), orange, mandarin
	2 ± 0.5, 14	Japan	De Lima *et al.* (2007), lemon
	2 ± 0.5, 16	Japan	De Lima *et al.* (2007), orange, mandarin
Rhagoletis mendax	0, 40	California	CDFA (2007), blueberries
Lepidoptera			
Cryptophlebia leucotreta	–0.6, 22	USA	APHIS (2007), stone fruit, grape
C. leucotreta	–0.6, 24	USA	APHIS (2007), citrus fruit
Conopomorpha sinensis	1, 17		
	1.4, 20	USA	APHIS (2007), carambola, lychee, longan, sand pear
Weevils (Coleoptera: Curculionidae)			
Conotrachelus nenuphar	0, 40	California	CDFA (2007), blueberries

[a] Temperatures originating from US regulations or research have been converted from Fahrenheit and rounded to nearest 0.1°C.
[b] In general, treatments for US markets are thought to satisfy 'probit 9' (99.9968%) efficacy requirements, while those for Japan provide at least 99.99% efficacy.

the Queensland fruit fly, *B. tryoni*, shows little difference in susceptibility within the range of 1–3°C (De Lima *et al.*, 2007). Jessup *et al.* (1998) found first instar Queensland fruit fly to be the most cold-tolerant stage in blueberries and that 12 days at 1°C prevented pupariation of over 400,000 treated insects.

Anastrepha spp., especially the Mexican fruit fly, *Anastrepha ludens*, seem to be more cold tolerant than Mediterranean fruit fly (Table 7.2 and 7.5). Benschoter (1984) found that three of 19,967 Caribbean fruit fly (*Anastrepha suspensa*) eggs and larvae pupariated after 7 days at 10°C followed by 17 days at 1.7°C and two adults emerged. The APHIS schedule of 17 days at 1.7°C for all *Anastrepha* spp.

other than *A. ludens* means that this treatment would fall short of being a 'probit 9' level (99.9968%) treatment for Caribbean fruit fly adult emergence let alone larval mortality.

Temperate tephritids require the longest cold treatment times; 40 days at 0°C is required for hosts of blueberry maggot, *Rhagoletis mendax* (CDFA, 2007).

Lepidoptera

A cold disinfestation treatment schedule against false codling moth, *Cryptophlebia leucotreta*, is −0.6°C for 22 or 24 days depending on the host fruit (APHIS, 2007). In June 2005, two interceptions of live larvae of this pest occurred in cold-treated clementines from South Africa to the USA. There were apparently no problems with the application of the treatment on board the two ships involved. In response APHIS added 2 days to the treatment for citrus fruit, making it now 24 days. Treatment time for stone fruit and grape was left at 22 days. Research done with this insect in the 1960s concluded that 24 days at −0.6°C would be necessary for quarantine security of the most tolerant stage, 2–3 day-old pupae (Myburgh and Bass, 1969). Another tortricid fruit pest of quarantine importance, lightbrown apple moth, *Epiphyas postvittana*, requires methyl bromide fumigation combined with post-treatment cold storage for 21 days at 0.6°C (APHIS, 2007).

Recently several live *Conopomorpha sinensis* (family Gracillariidae) were found in cold-treated imported lychee leading APHIS to revise the cold treatment schedule for this fruit.

Weevils

Compared to Diptera and Lepidoptera, cold disinfestation is utilized against few Coleoptera pests of fruit and vegetables. The plum curculio, *Conotrachelus nenuphar*, is one example; 0°C for 40 days is required by the state of California (CDFA, 2007). Potential exists for utilization of cold against other coleopterous pests, and this possibility deserves further research.

Mites

Mites are a commonly occurring quarantine pest of fruit and cut flowers. APHIS (2007) includes the oriental citrus mite, *Eutetranychus orientalis*, in its cold schedules. Other species for which cold might also be an option have overwintering stages which could be highly cold tolerant. In commodities where these species are absent cold justifies investigation as an option.

Tetranychid mites are an actionable pest when found at inspection on imports of fruit and flowers to Japan and exporters are specifically advised to ensure that they are not present (Anon., 1989). Cold storage, where appropriate, would make a beneficial contribution towards avoiding interception and subsequent rejection or disinfestation action which could affect product quality.

Selection of Cold as a Treatment Option for Perishable Commodities

There are three options for determining when treatment is done: pre-shipment, in transit and on arrival. If the shipping transit time is less than the treatment duration, pre-shipment treatment is normally the most favourable option. Pre-shipment treatment is almost invariably necessary where the importing country requires pre-shipment supervision of treatment and pre-clearance by an inspector of its own regulatory authority. In-transit treatment is effectively restricted to sea transport. It is approved by the USA (APHIS, 2007) and other countries provided that strict conditions are met. These conditions include continuous temperature monitoring and recording for both the commodity and the air within the ship's hold or the container (Fig. 7.1). The temperatures required and precision of calibration and control differ from regular in-transit cool storage for quality purposes so special equipment is usually needed.

An advantage common to all cold treatment is that packaging can be done before treatment. If forced-air cooling is used the packages need vents at each end to permit the airflow needed for the more rapid cooling that this method can give. These holes are sometimes considered to be a post-treatment risk of re-infestation and may be required to be gauzed. However, in reality the risk of re-infestation is not significant for most pests especially those held and transported cold.

Treatment on arrival is usually only an option for low risk commodities but where possible it allows the importing country complete control over the process

Fig. 7.1. A refrigerated shipping container capable of use for in-transit quarantine disinfestation treatment at a temperature precision of ± 0.5°C. The power source is mains power from the ship's supply or a land source. A data recorder (upper right of refrigeration unit) provides a charted history of temperatures over the transit time at a number of locations within the container.

and can ensure reliability of application in line with the perceived quarantine security requirement.

Durable Commodities including Grains and Seeds

Cold treatment has an increasingly important role to play in the treatment of durable commodities for phytosanitary purposes. With the exception of khapra beetle, *Trogoderma granarium*, and a very few other pests, most pests of stored grain and other durable commodities except timber are cosmopolitan in distribution. This excludes them from categorization as quarantine pests. However, phytosanitary action against them, specifically and collectively under the terms of legislation of exporting and importing countries may be justified under the category of regulated non-quarantine pests. The level of action required does not normally involve prophylactic treatment as for fruit flies but rather action to be taken should they be detected in pre-shipment or point of entry inspections.

Commercial considerations have resulted in a move away from point-of-export inspection treatment and treatment in response to detection, to avoidance of detectable infestations by pest management in longer term pre-export storages (Kelly and Wilkin, 1994). This involves available means of pest management including aeration with ambient or refrigerated air. In temperate parts of the world, winter temperatures are such that ambient aeration will disinfest large bulks of grain at a high level of efficacy. In general, temperatures below 15°C will prevent population increase in most pests and temperatures below 8–10°C will eventually disinfest the bulk of arthropod pests. Elsewhere, smaller volumes can be treated with refrigerated air (Fig. 7.2) and for small quantities of valuable

Fig. 7.2. A mobile refrigerated aeration unit suitable for small bin lots such as seeds.

commodities such as seeds, deep-freezing to $< -18°C$ should find ready acceptance as a quarantine treatment against insect pests.

Ambient aeration of a dry commodity such as stored grain or seed requires special controlling mechanisms to ensure that only appropriately cold dry air is used so that the moisture content of the commodity is not unacceptably raised. This requires engineering expertise and the use of time-proportioning or wet bulb temperature controllers.

Future Considerations

Cold is one of the oldest and most used phytosanitary treatments, and prospects for future use are favourable. Research with cold should continue to include more quarantine pests and modifications in the treatment to preserve quality of commodities that do not tolerate cold well (Nishijima *et al.*, 1995). Some old treatments for which supporting data do not seem to be readily available should be re-examined in the light of the apparent failure of cold treatments against Mediterranean fruit fly in clementines from Spain.

Some kohl and other vegetables are additional possibilities for cold disinfestation, depending on the pest species involved. Fruits and vegetables for which low oxygen storage at low temperatures is normal practice may be disinfested of many pests as a consequence. Chapter 11 covers modified atmospheres and the effects of combining cold with them.

One area in which cold is largely unresearched is cut flowers and foliage (Mangan and Hallman, 1998), although many of these commodities tolerate the cold temperatures probably needed to kill their quarantine pests (Williams *et al.*, 2005).

8 Phytosanitary Heat Treatments

Heating commodities to temperatures lethal to pests without damage to the commodity can be accomplished by a variety of techniques. In the recent past, more disinfestation research effort has been expended on heat treatments than any other. For example, the *Journal of Economic Entomology* has a section on phytosanitary entomology and 36% of the 22 articles published during 2004–2005 involved heat as the primary control factor. The next largest category was ionizing irradiation at 22%. The rest of the research effort was expended on fumigation (17%), cold (11%), modified atmospheres (8%) and non-host status (6%).

Using heat to disinfest fresh fruits, vegetables and ornamental plants is more challenging than its use on durable commodities because the fresh commodities are also living, respiring organisms vulnerable to heat injury. In practice, heat is one of the most challenging areas of phytosanitary treatment research and commercial application because of the number of variables that affect efficacy and fresh commodity quality and the difficulties in attempting to apply generic heat treatments across pest species and commodities.

A new volume on postharvest heat treatments (Tang *et al.*, 2007) reviews heat transfer theory, modelling, control of postharvest decay organisms, the influence of heat shock proteins, treatment protocol development, commercial treatments, heat disinfestation of structures, and effect of heat disinfestation on quality of fresh commodities among other topics. It was not printed by the time writing on this book closed, so we cannot offer details of it here. Other recent reviews of heat treatments include factors affecting phytosanitary treatment efficacy (Hallman, 2000) and physiological responses of insects to heat disinfestation treatments (Neven, 2003).

There have been some noteworthy failures of phytosanitary heat treatments as a result of the effect of biological and physical variations in treatment parameters. Unlike other major treatments (cold, fumigation and irradiation) heat treatments are not applied in a generic sense. Variations in the shape and size of the commodity being treated as well as the species of pest often require that a specific treatment be done for a specific pest-commodity combination. For example, the time required to disinfest mangoes of tephritid fruit flies using hot water immersion depends on the weight, shape and origin of the mangoes (APHIS, 2007). In contrast, cold and methyl bromide fumigation treatments can be applied to a number of fruits regardless of size (FAO, 1984; APHIS, 2007) and all hosts of all tephritid fruit flies in all countries may be disinfested by irradiation at 150 Gy (APHIS, 2007; IPPC, 2007). Nevertheless, the lethal time and temperature values that are physiologically based for heat might be found eventually to be no more variable than generic values for cold and irradiation. Such information could then be used to formulate generic phytosanitary heat treatments across host commodities and groups of pests.

Commercially available heat treatments fall into two general categories depending on the carrier of the heat, air or water. Water is a more efficient and uniform carrier of heat compared with air (Stewart *et al.*, 1990), although heated air is more broadly applicable than water as the commodity being treated does not need to be wetted and fresh commodities may tolerate heated air better than heated water. Heated-air or water treatments that kill quarantine pests without unacceptably injuring commodities are now widely available.

Unlike cold treatment, gamma irradiation or fumigation, heat treatments are typically applied before packaging. A mode of treatment of commodity in cartons analogous to forced-air cooling should be possible theoretically. However, it might be difficult to implement in commercial practice owing to the relatively short disinfestation treatment times, for heat compared with cold, tolerated by most commodities. Where there is a perceived lower risk of product injury with heated-air treatment compared to hot water dipping, it can be the preferred method of disinfestations and is the preferred method of disinfestation for most fruit and vegetable hosts of endemic tephritid fruit flies in the Asian Pacific regions. Temperatures for heated-air treatments on fresh commodities against tephritid species in general vary from 43 to 52°C.

Thermotolerance of the commodity is usually the factor dictating treatment temperature. Treatment times to ensure no survivors at temperatures below 43°C are generally considered too long, although this does not rule out circumstances that might favour a 24 h heat treatment, especially if commodity quality problems could be resolved. Fresh commodities generally do not tolerate temperatures above 52°C for the time periods required to kill pests. Moss and Chan (1993), reporting on a study of thermal death kinetics between 37 and 50°C of embryos of Caribbean fruit fly, *Anastrepha suspensa*, suggested that perturbation of the cell membrane or a macromolecule may occur between 42 and 43°C greatly accelerating thermal death rates above that temperature. Keeping treatment times as short as possible is usually preferred by industry in order to prevent any backlog of commodity to be treated. Shorter treatment times generally result in savings in treatment and handling costs.

It must be noted that much research reported to tenths or hundredths of a degree was done using whole degree increments in the Fahrenheit scale (which still is the temperature scale used commercially in the USA) and converted to Celsius for publication and that high precision may not necessarily be inferred. Some research prior to about 1980 was not done with the level of precision justifying reporting to even tenths of a degree.

Heated-air Treatments

Transfer of heat to the commodity with heated-air treatments is by convection (air to fruit surface) and conduction (fruit pulp to pulp). Japan, USA and New Zealand have each developed commercial forced-air treatment systems primarily for the treatment of fruit against tephritid fruit flies (Figs 8.1 and 8.2). A variety of heated-air treatments have been used and researched for phytosanitary control, especially against tephritid fruit flies (Table 8.1). The terminology for specific heated-air treatments (dry heat, steam pressure sterilization, vapour heat, moist heat, high temperature forced-air, hot air) may be inconsistently or incorrectly used in reports and treatment protocols (Hallman and Armstrong, 1994). Therefore, it is important to give a complete description of how the treatment is applied instead of relying on a treatment name to convey treatment description.

The efficacy of heated-air treatments is affected by several variables besides the main ones (temperature and time) such as humidity, air speed, airflow direction, headspace in the chamber, oxygen concentration in the chamber, design of the treatment chamber, as well as size, shape and orientation of the individual commodities being treated (Mangan and Hallman, 1998). The capacity of air to carry heat is directly related to moisture content; moist air will heat products faster than dry air.

Because of the relatively high latent heat of condensation of water (2272 J/g) saturated air should theoretically heat commodities as fast as hot water immersion, and this has been substantiated at least on a small scale (Shellie and Mangan, 2000). However, for practical purposes, as saturated air passes through a commercial commodity load, water vapour condenses leaving less to condense

Fig. 8.1. A double-unit commercial forced-hot-air treatment facility showing the entry doors to the treatment chamber, a pallet that enables relatively unrestricted airflow, control and recording equipment cabinet and the steam generator for humidification.

Fig. 8.2. Inside a forced-hot-air facility similar to that in Fig. 8.1 showing fans for the circulation of air down through carriers with perforated bottoms. Entry is from the far side and egress from the open side to a secure packing area.

Table 8.1. Examples of heated-air phytosanitary schedules against tephritid fruit flies.

Pest	Commodity	Temperature (°C)	Relative humidity (%)	Time	Reference
Anastrepha suspensa	Carambola	43.5–46.5	~100	1–2 h including approach time	Hallman (1990)
Ceratitis capitata	Papaya	46–47	40–60	5 h including approach time	Armstrong *et al.* (1989)
Bactrocera cucurbitae	Momordica	45	95–98	30 min plus approach time	Sunagawa *et al.* (1988)
Bactrocera cucumis	Zucchini	45	> 94	45 min plus approach time	Corcoran *et al.* (1993)
Bactrocera dorsalis	Mango	46.5	~97	10 min plus 2 h approach time	Unahawutti *et al.* (1992)
Bactrocera tryoni	Tomato	44	92	2 h plus approach time	Heather *et al.* (2002)

further down the air stream, resulting in less than maximum possible heating of the entire load.

The same effect reduces the theoretical efficiency of all components of heated air. Not only does water vapour lose heat as it travels though a commodity load so does oxygen and nitrogen, the other major components of air. Therefore, when heated air is scaled up to commercial volumes the treatment usually takes longer to apply than it does under small-scale, experimental conditions. Because of this it may be important for researchers to slow down the experimental heating rate of heated-air treatments to approach expected commercial conditions more closely as heating rate can significantly affect efficacy and commodity quality (Paull and McDonald, 1994; Neven, 1998).

Factors that affect treatment efficacy also affect commodity quality. Adequate moisture preserves commodity quality. Low levels of moisture will dry fresh commodities and high levels may cause condensation on the commodity, which is sometimes thought to reduce commodity quality (Shellie and Mangan, 2000). The most favourable type of heated-air treatment for maintaining quality of fresh commodities and providing reasonable pest-kill times appears to include keeping the dew point in the chamber as close to, but below, the surface temperature of the commodity to allow for the high humidity content to carry more heat than drier air and prevent the commodity from losing moisture (Fig. 8.3). Keeping the dew point below the surface temperature of the commodity prevents water condensation on the commodity which might interfere with gaseous exchange between the interior of the commodity and the atmosphere and promote injury. Furthermore, if the commodity was wet at the end of the treatment it may have to be dried before further processing.

An actively circulating airflow will keep treatment times and temperature variations to a minimum. Typical core-pulp treatment temperature requirements range from 43 to 48°C depending on run-up and holding times (Table 8.1) and need to be monitored by sensing probes inserted into the flesh (Fig. 8.4). If the circulation system is very efficient, a differential as small as 1°C is possible between the air and the required core temperature even for large individual commodity components (Fig. 8.3). This has obvious benefits for avoidance of product injury while permitting the highest possible lethal temperature. Air circulation is critical, with small differences in the arrangement or load of the commodity within the treatment chamber liable to affect disinfestation efficacy (Mangan and Ingle, 1992; Heather *et al.*, 1997).

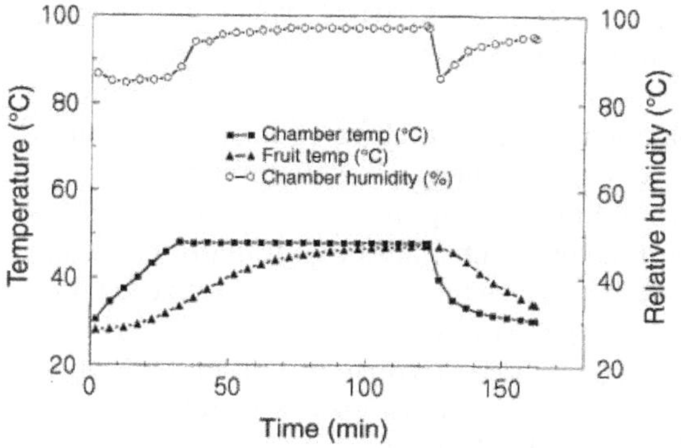

Fig. 8.3. Temperature and humidity profiles for a forced-hot-air treatment to disinfest mangoes against *Bactrocera tryoni* involving a fruit core temperature of 47°C for 15 min followed by hydrocooling (Source: Heather *et al.*, 1997; reprinted by courtesy of Elsevier BV).

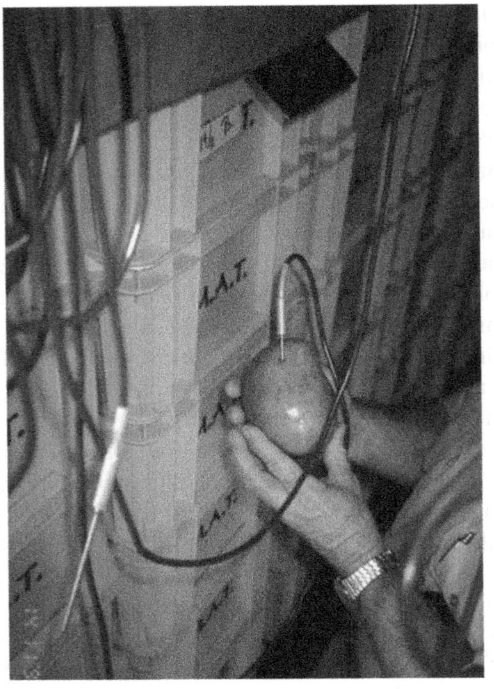

Fig. 8.4. Resistance temperature detector probes used to control and monitor fruit treatment in a forced-air heating unit. The probe monitoring fruit pulp temperature would be one of several distributed throughout the load, each inserted into the largest fruit of the plastic basket. (Photograph by R. Jordan)

Hallman and Armstrong (1994) provide a history of heated-air treatments from their beginnings in the 1920s through to the early 1990s. Heated-air treatments began with a room with a commodity in field boxes filled with a circulating and replenishing supply of heated air and vapour, resulting in a relative humidity (RH) of about 100%. This initial treatment was called 'vapour heat' and required up to > 16 h to complete. Temperatures were gradually raised over several hours until they reached about 43°C and not allowed to rise any higher in part because temperature control devices and heaters could not be trusted to prevent lengthy heat spikes that might damage the fruit. A blower kept the air moving and flowing through the commodity load. Water usually condensed on the commodity, although it might dry again before the treatment was finished as the surface temperature rose to exceed the dew point. It was used successfully to treat citrus fruit in Florida for export during the first Mediterranean fruit fly, *Ceratitis capitata*, infestation in that state in the 1920s and later in Texas and Mexico against Mexican fruit fly, *Anastrepha ludens*, and other species until the early 1950s when fumigants replaced heat.

Time of treatment was eventually halved by a more abrupt increase in temperature, although this sometimes resulted in damage to some fresh commodities. Some treatments, such as to Hawaiian papayas, began using less than 100% RH, which seemed to maintain commodity quality better than saturation. By the 1950s higher treatment temperatures (up to 49°C) were being studied as improved equipment made temperature measurement more accurate and stability more achievable. When commodity disinfestation began to be done by fumigation the latter's ease of use and cheap cost largely replaced heated-air treatment of fresh commodities.

Two factors led to a revival of phytosanitary treatments using heated air. The first was that by the late 1970s research with heated-air treatments commenced in Japan as part of a campaign to eradicate and maintain area freedom from Oriental fruit fly, *Bactrocera dorsalis*, and melon fly, *Bactrocera cucurbitae*, throughout the islands of Japan. Because solanaceous fruits, including bell peppers and aubergines to be shipped from Okinawa to the main islands of Japan, did not tolerate ethylene dibromide or methyl bromide fumigation well, a treatment was developed to disinfest both fruits against *B. dorsalis*, by heating with 43.9 ± 0.3°C air at > 90% RH until the centre of the fruits reached 43°C and then holding for an additional 3 h before cooling (Sugimoto *et al.*, 1983). Subsequently, additional heated-air treatments were developed to treat mangoes from the Philippines, Thailand and Taiwan for export to Japan and other fruits including papaya from Hawaii against *B. dorsalis* and *B. cucurbitae* (Sunagawa *et al.*, 1988; Unahawutti *et al.*, 1992).

The second factor leading to a revival in heated-air treatment research was the loss of ethylene dibromide in 1984 in the USA because the fumigant was considered a carcinogenic and mutagenic risk. Although methyl bromide returned to replace some of the phytosanitary uses of ethylene dibromide (which were usurped by ethylene dibromide 30 years earlier) non-chemical alternatives, including heated air, were actively sought especially when methyl bromide was implicated as a stratospheric ozone-depleting substance in the early 1990s. Heated-air treatment units were developed that were improvements over older designs in that airflow was forced through the commodity load (before then the airflow pattern was somewhat open), temperature and humidity measurements were refined, and heat and humidity controls were more responsive (Williamson and Winkelman, 1989; Gaffney and Armstrong, 1990; Gaffney *et al.*, 1990).

One of the highest temperature heat treatments resulting from this round of research is for mangoes from Mexico against *Anastrepha* spp. fruit flies. Air at 50°C is introduced straightaway into the chamber and the treatment is concluded when the mango seed surface reaches 48°C (APHIS, 2007). Treatment times for these new heated-air treatments are about one-third of those required by the older, lower temperature 'vapour heat' treatments.

Modified Atmosphere and Heat

Reduction in oxygen concentration and/or increase in carbon dioxide or other atmospheric constituents comprise a class of phytosanitary treatments called modified atmospheres. Temperature is positively correlated with mortality in modified atmosphere treatments. This fact has been exploited to develop heat/modified atmosphere treatments that achieve control in less time than either treatment by itself (Neven and Rehfield-Ray, 2006; Neven *et al.*, 2006b). This topic is further covered in the chapter on modified atmospheres (Chapter 11).

Dry Heated-air Treatments

There is anecdotal evidence that dry heat has been used to disinfest grain since early historical times. Humidity is not controlled for dry heated-air treatments. These treatments are mainly for durable commodities as the treatment temperature is usually above the maximum limit for fresh commodities, which is about 52°C. For example, 85°C for 4–12 h is used to disinfest broomcorn, dried plants, peat and sphagnum moss, chestnuts and rice straw and 95°C for 48 h is used to devitalize bird seed (FAO, 1984). At the temperature extremes of dry heat, bagging materials are treated at about 54°C for 1 h while herbs may be treated at as high as 120°C for 2 h. One of the few disinfestation possibilities for dry heat on a fresh commodity is 39.4°C for 30 h for sweet potato roots to disinfest them of sweetpotato weevil, *Cylas formicarius elegantulus* (FAO, 1984).

For grain, a fluidized bed heating system for disinfestation was developed as an alternative to pesticides to which pests had developed resistance (Dermott and Evans, 1977). This system was designed to enable disinfestation of grain flows at rates applicable to loading on to large bulk carrier ships. The fluidized bed technology enabled individual grains to be heated rapidly to around 60°C using hot air of up to 100°C at high pressure in a reverse direction to the grain flow. Time-temperature regimes are based on the $LD_{99.9}$ for the lesser grain borer, *Rhyzopertha dominica*, the grain pest with the highest known heat tolerance (Evans and Dermot, 1981). The baking quality of grain was unaffected at these temperatures. A further use of heat on grain is for drying with hot air. Depending on the temperatures reached, significant pest mortality could be expected.

Heating wood packaging material to a minimum core temperature of 56°C for 30 min controls wood-boring nematodes and insects as detailed in ISPM 12 (IPPC, 2007). Ash (*Fraxinus* sp.) logs may be disinfested of emerald ash borer, with any heat source (e.g. air, steam kiln, hot water) that raises log centre temperature to a minimum of 71°C for 75 min (APHIS, 2007).

Steam Treatments

Steam at normal atmospheric pressure (≥ 100°C) will not only kill arthropods in a short period of time but will also kill pathogens except some spore stages in hardy materials, such as dunnage, packaging, pallets and rice straw mats. To kill spores requires steam treatment under pressure which enables higher temperatures to be achieved. For example, steam under pressure (134°C) for 4 h is used to disinfest bagging materials (FAO, 1984).

Early in the 20th century in Australia heating to 58–60°C for 3 min was done in bins using piped steam to disinfest large stocks of Australian grain when overseas exports were suspended for some years due to war. The process was highly labour intensive as grain then was stored bagged.

A novel use of steam as a phytosanitary treatment is to alternate it with vacuum. Steam-vacuum cycles lasting 0.1–0.5 s killed the surface pest red scale, *Aonidiella aurantii*, on lemons, although the treatment may have predisposed some of the fruit to accelerated decomposition (Fuester *et al.*, 2004).

Hot Water Treatments

Sharp (1994) reviewed the use of hot water immersion as a phytosanitary treatment. Like the revival of heated-air treatments in the 1980s, the development of hot water immersion as a phytosanitary treatment was in direct response to the loss of ethylene dibromide as a fumigant. Armstrong (1982) first showed that hot water immersion alone could be used to disinfest fruit against fruit fly eggs and larvae. Immersing bananas in water at 50°C for 20 min prevented an estimated total of > 222,000 eggs of three species of tephritids in the fruit from reaching the puparial stage.

Armstrong (1982) was inspired in his research by Seo *et al.* (1972) who developed a complex three-component method for disinfesting mangoes of tephritid fruit flies. The method consisted sequentially of immersion in 46.3°C water for 20 min, fumigation with 8–12 g/m^3 ethylene dibromide for 2 h at 21°C, and finally refrigeration at 7.6°C for 4 days. The 20 min hot water immersion by itself gave 82% mortality of Oriental fruit fly. They used this combination because the 20-min hot water immersion was being used to control surface infection by the anthracnose fungal pathogen on mango (Smoot and Segall, 1963). Ripe mangoes were typically stored at 7.2°C, and it was apparently thought that some combination of the two sequentially or either treatment separately would be insufficient, thus necessitating a supplementary fumigation. The dosage of ethylene dibromide used, 8–12 g/m^3 at 21°C, was lower than that required as a stand-alone treatment for tephritids, which was at least 15 g/m^3 (FAO, 1984). We do not know if the treatment developed by Seo *et al.* (1972) was ever used commercially.

One of the earliest commercially used ethylene dibromide replacements was a two-part hot-water immersion treatment for papaya in Hawaii (Couey and Hayes, 1986). Papaya fruit less than one-quarter ripe were immersed in 42°C water for 30 min followed immediately by immersion in 49°C water for 20 min. After the heat treatment these fruit were immediately cooled with an ambient water spray in order to minimize injury. The treatment was designed to kill only eggs and peripherally located first instars; it would not provide complete kill of later instars deeper in the fruit. Approval was only for papayas less than one-quarter ripe to preclude the possibility of larvae beyond the first instar. However, soon after commercial implementation live third instar oriental fruit flies were found in papayas treated at less than one-quarter ripe (Zee *et al.*, 1989). Further investigation discovered a so-called 'blossom end defect' whereby some papayas had a small opening from the blossom end into the cavity in the fruit. Apparently the Oriental fruit fly oviposited earlier at these openings than expected, resulting in later instars inside the fruit that were not controlled by the relatively mild double-dip heat treatment.

When the initial research was conducted, Couey and Hayes (1986) found six surviving larvae but dismissed them as 'accidental re-infestation'. In hindsight, they may not have been. Also the large-scale testing conducted during the confirmatory part of the research used naturally infested fruit from one orchard. It was not recorded to what level these fruit might have been infested, thus there was no estimate of the degree of rigour that the test provided. The opening at the blossom end was probably present during the research phase of the treatment's development but was not noticed (Mangan and Hallman, 1998).

Phytosanitary treatment research is often conducted on an ad hoc basis. For example, a long-standing treatment could be restricted (ethylene dibromide and methyl bromide fumigations) leaving no ready alternative; or a treatment that has been thought to be effective fails (cold treatment against Mediterranean fruit fly, hot water immersion of mangoes, double-dip hot water treatment of papayas) requiring urgent research to reopen trade as soon as possible. This type of research situation can lead to erroneous assumptions, short cuts and mistakes. It would be wiser if countries relying on phytosanitary treatments had alternative measures in reserve for important commodities. Also, key treatments need to be reviewed periodically for potential problems so as to avoid them or be ready for them when they appear, as seems inevitable with phytosanitation.

After considerable research, hot water immersion was used commercially to disinfest mangoes of Mediterranean fruit fly and *Anastrepha* spp. in the late 1980s (Sharp, 1994). In 2000, live larvae were found in the USA in hot-water treated mangoes imported from Mexico (Scruton, 2000). As a consequence treatment times were extended and the industry practice of immediate cooling after treatment was proscribed.

The mango incident highlights the consequences when there is a disconnect between research and its application. Finding live larvae upon inspection of heat-treated fruit implies a failure of the treatment process. However, in the experimental procedure used to develop the mango hot-water treatment, treated mangoes were put in larval collection towers and the larvae therein were not counted as surviving the treatment unless they emerged from the fruit on their own or were washed out after some weeks and pupariated normally (Sharp, 1988). An unknown percentage of those larvae remaining in fruit may have been alive for some time after the treatment, but failed to emerge from the fruit. Although these larvae were not counted as survivors by researchers, regulators may find them at an early post-treatment inspection and incorrectly judge the treatment inefficacious.

However, not only was there a disconnect between research and regulation in the case of hot water immersion of mangoes, but the initial research was not as rigorous as it should have been. 'Normally formed' puparia were considered survivors, although a definition of puparial normality and its limits was not provided (Sharp, 1988). It was assumed that adults would not emerge from 'abnormal' puparia. Thomas and Mangan (1995) identified two morphologically abnormal types of puparia after hot water immersion of third instars of two species of *Anastrepha* that infest mangoes. Adult emergence rate was high for a class dubbed 'bottlenosed' puparia, and low for larviform puparia, showing that adults may emerge from malformed puparia.

However, the tribulations of hot water treatment of mangoes may not be over. Indeed, one can never be sure that a phytosanitary treatment will not become challenged as factors including the pests themselves change in the future. In a study to correct the deficiencies of the hot water treatment of mangoes Shellie and Mangan (2002) tested two types of mango infestation techniques looking for the most appropriate one to test efficacy of the treatment on the fruit. One technique (cage) was closer to feral infestation of mangoes in that the fruit were placed in a cage with adult fruit flies and resulting larvae allowed to develop to the desired stage for treatment. The other (artificial) relied on placement of larvae reared in diet in to holes bored through the pulp to the seed surface in mangoes and sealed with a plug of mango pulp and peel with hot-melt glue. After hot water immersion at 46.1°C for 70 min, survival of third instars, in cage and artificially infested mangoes was 42 and 3.7%, respectively, indicating that larvae from the cage technique were harder to kill with heat. Furthermore, heating of the cage-infested mangoes occurred at a faster rate, reaching almost 4°C higher temperature at the seed surface at the end of the treatment. Unfortunately, the artificial infestation technique was used in the research confirming new hot-water treatment schedules for mangoes. Therefore, the stage may be set for further failures in hot-water immersion treatment of mangoes.

Hot water immersion of mangoes has the dubious distinction of being the only treatment known to have resulted in human deaths because of consumption of fruit treated for phytosanitary purposes. Fruit expands slightly as it is being heated and may absorb water from the treatment tank. The water is supposed to be adequately treated with chlorine to kill any pathogens, but in at least one instance hot-water treated mangoes from a farm were the common denominator in a broad case of *Salmonella* infection that resulted in at least 78 sicknesses and two deaths from 13 US states (Penteado *et al.*, 2004).

Hot water immersion is the principal phytosanitary treatment for mangoes produced on the two American continents. Nearly all of more than 220,000 t of mangoes imported into the USA each year are disinfested with hot water at 46°C for 65–110 min (Fig. 8.5). Treatment time depends on the weight, shape and origin of the mangoes (APHIS, 2007). Hot water immersion has been studied for a number of other commodities and applied commercially (49°C for 20 min) to limes for mealybugs and other surface pests and lychee and longans for tephritid fruit flies (Table 8.2).

Hot water immersion is approved to disinfest plant propagative materials, such as bulbs of snails; chrysanthemum cuttings of leafminers, surface insects and mites; orchids of leafminers and *Eurytoma* sp. wasps; and narcissus of mites (APHIS, 2007). Heat may reduce vigour of some propagative materials, even if the treatment has been approved by regulatory agencies, and should be checked before being applied on a large scale.

Hot water, having a high surface heat transfer coefficient, is more efficient at transferring heat than heated air (Stewart *et al.*, 1990). Water has a high volumetric heat capacity and is easily moved, so it surrounds each individual in a commodity load with uniform heat if properly agitated. Hot water immersion requires precise temperature control if injury is to be avoided to susceptible commodities. For example, the minimum hot-water treatment temperature for

Fig 8.5. A hot-water treatment unit for disinfesting fruit in palletized batches. Note the active circulation of the water.

Table 8.2. Examples of hot-water disinfestation schedules for fresh perishable commodities.

Pest	Commodity	Temperature (°C)	Time	Reference
Diptera				
Anastrepha fraterculus	Mango	46	1.5 h	Sharp and Picho-Martinez (1990)
Anastrepha suspensa	Guava	46.1 ± 0.5	35 min	Gould and Sharp (1992)
Coleoptera				
Asynonychus godmani	Lemon	52	8 min	Soderstrom *et al.* (1993)
Homoptera				
Pseudaulacaspis cockerelli	Ornamental (*Strelitzia reginae*)	49	5–6 min	Hara *et al.* (1993)
Lepidopetera				
Cryptophlebia spp.	Lychee and longan	49	20 min	Follett and Sanxter (2001)

longan is 49°C and injury may occur if the temperature exceeds 49.5°C (APHIS, 2007). In common with other heat treatments it is done before packaging so subsequent quarantine security procedures need to be established. Many commodities float, so a method for keeping them underwater must be devised for uniform heating. Depth may be dictated; APHIS (2007) uses a standard 10.2 cm (4 inches) for the minimum distance between the top of the load and the water surface for hot water treatments of fruit.

Hot water immersion is easier and cheaper to scale up to commercial capability than other heat treatments and is usually done as a batch treatment. It has been attempted on mangoes in a conveyor system but fruit damage occurred, probably because the mechanical tumbling of the mangoes during heating caused bruises. Heating while bruising fruit may accentuate injury symptoms. Because hot water immersion is not as susceptible as heated air to factors that affect heat transfer, a single temperature-time schedule (e.g. lychee in water at 49°C for 20 min; APHIS, 2007) is usually specified for hot water immersion that does not depend on the temperature reached in the core of individual pieces while the commodity is being treated. The time required to ensure mortality of any pests present will have been determined empirically. The initial minimum core temperature of the commodity should be specified so that the commodity reaches lethal temperatures by the end of the treatment: for example mangoes must be at ⩾ 21.1°C before hot water treatment is initiated (APHIS, 2007).

Water temperature can be maintained constant throughout the immersion tank to within 0.5°C using a high capacity heat source such as reticulated steam, gas, oil or electricity and high capacity circulation pumps (Fig. 8.5). Hot water treatments are very specific; indeed, they are probably the most specific of all the commercial phytosanitary treatments as evidenced by multiple treatment schedules for mangoes based on weight, shape and origin (Table 8.3). The specificity is because treatments do not depend on a temperature end point. Weight, shape and density of the commodity affect the time required for lethal heat to reach the centre of commodities. If the end point of hot water immersion was a specific core temperature, as is usually the case for heated-air treatments, hot-water immersion treatments could probably be made more generic.

Postharvest hot water dips and drenches (hot water poured over or flood-sprayed on commodities) are used against incipient fungal pathogen infections in horticultural crops (Lurie, 1998). These surface treatments, typically at 51–55°C for 5–15 min against pathogens would not have high levels of efficacy against fruit fly larvae within the pulp of fruit but have been found to be effective against surface pests such as mealybugs (Hara and Jacobsen, 2005). Conversely, longer-timed fruit fly disinfestation immersion and other heat treatments, albeit at lower temperatures, might have a degree of therapeutic effect against the pathogens (Jacobi *et al.*, 1994).

Despite the problems with efficacy, commodity quality and even human health that hot water treatments have had, it remains a viable treatment that should be investigated for other commodities because of its broad efficacy against pests, relative low cost and simple ease of application. Problems with commodity tolerance of hot water immersion may be ameliorated to some extent through gradual heating of the water and commodity (McGuire, 1991) or preconditioning of the commodity with a sublethal warm temperature before the actual heat treatment (Jacobi *et al.*, 2001).

Table 8.3. Immersion times in 46.1°C water for mangoes imported by the USA depending on origin, shape and weight (Source: APHIS, 2007).

Origin	Shape	Weight (kg)	Time (min)
Puerto Rico, US Virgin Islands and northern West Indies	Flat, elongated	Up to 0.4	65
Puerto Rico, US Virgin Islands and northern West Indies	Flat, elongated	0.4–0.57	75
Puerto Rico, US Virgin Islands and northern West Indies	Rounded	Up to 0.5	75
Puerto Rico, US Virgin Islands and northern West Indies	Rounded	0.5–0.7	90
Puerto Rico, US Virgin Islands and northern West Indies	Rounded	0.7–0.9	110
Mexico, Central and South America and southern West Indies	Flat, elongated	Up to 0.375	65
Mexico, Central and South America and southern West Indies	Flat, elongated	0.375–0.57	75
Mexico and Central America less Panama	Rounded	Up to 0.5	75
Mexico and Central America less Panama	Rounded	0.5–0.7	90
Mexico and Central America less Panama	Rounded	0.7–0.9	110
South America and southern West Indies	Rounded	Up to 0.425	75
South America and southern West Indies	Rounded	0.425–0.65	90

When considering modifications of heat treatments to prevent damage to commodities, any reduced efficacy against the pest must also be tested and remedied if it occurs. For example, Yin *et al.* (2006) found that exposing fifth instar codling moth, *Cydia pomonella*, to a non-lethal 35°C for times ranging from 40 to 1080 min significantly increased the insect's tolerance to subsequent exposures to 48–52°C.

It is hypothesized that hot water immersion may be more damaging to fresh commodities than heated air. Hot water imparts a damaging thermal shock to commodities suddenly immersed in it and there is depletion of oxygen in the hot water during the treatment that results in stressful anaerobic conditions (Hayes, 1994; Bollen and Dela Rue, 1999). However, Shellie and Mangan (2000) concluded that hypoxic conditions in hot-water-immersed fruit were not responsible for increased injury compared with heated-air treatments. Hot water drenching should be investigated as a phytosanitary treatment if it causes less damage to fresh commodities although care will need to be taken to avoid channelling of the water flow which could result in uneven heating.

Electromagnetic Heat Treatments

Products can be heated by exposing them to electromagnetic processes that result in heat being produced in the product. Two are discussed here, one that has received considerable phytosanitary research attention, radiofrequency heating, and one that has not, ohmic heating.

Radiofrequency heating

The attractiveness of using radiofrequency (RF) heating as a phytosanitary treatment is based on three hypotheses: (i) commodities will heat more uniformly throughout, thus resulting in less peel injury compared with hot water and heated air, both of which treatments over-expose the peel in order to get sufficient lethal heat to the interior of a commodity; (ii) RF treatments will be much faster than other forms of heating; and (iii) it might be possible to target pests in commodities with RF heat without heating the surrounding commodity as much. Despite the fact that considerable research has been dedicated to RF heating as both phytosanitation and to kill insects in stored products since its insecticidal use was first proposed in 1928, no commercial usage has developed, and the three abovementioned hypotheses do not seem to hold very well.

Hallman and Sharp (1994) summarized research on RF heating to kill insects through the early 1990s and concluded that, although the use of RF heating as a phytosanitary treatment for fresh commodities did not look promising, it may have potential for dried and stored-product commodities. A more recent and upbeat review of RF heating for phytosanitation is by Tang *et al.* (2007).

RF or dielectric heating happens when radio waves are passed through a substance causing polar molecules, such as water, to rotate rapidly, causing heat from friction with neighbouring molecules. Non-thermal lethal effects, sometimes claimed, have not been substantiated (Hallman and Sharp, 1994). Some authors make a distinction between RF heating and microwave heating. The RF spectrum spans about 3 Hz to 300 GHz, and microwaves constitute the high frequency end of this spectrum (i.e. about 300 MHz to 300 GHz). Boundaries between the different parts of the electromagnetic radiation spectrum are somewhat arbitrary.

By the turn of the 21st century a revival of research with RF heating as a phytosanitary treatment was initiated by a group in the Department of Biological Systems Engineering at Washington State University in Pullman, colleagues at the University of California at Davis, the Agricultural Research Service (ARS) of the US Department of Agriculture (USDA) and the manufacturer of RF heating equipment Strayfield Fastran in Berkshire, UK. This consortium has the resources to develop a commercial RF phytosanitary heating system, if it is at all feasible.

The recent evolution of RF phytosanitary research on fresh fruit found that the treatment should be conducted in water to prevent arcing between adjacent pieces of fruit and overheating of the peel. The water should have similar dielectric properties as the commodity to be heated to achieve a similar heating rate for water, fruit peel and pulp (Ikediala *et al.*, 2002; Wang *et al.*, 2003). The product should be in motion to achieve more uniform heating (Birla *et al.*, 2004). Research into RF heating then focused on intermittent and slower heating to allow the load to reach more homogenous temperatures throughout. In fact, heating times have approached times that might be required to do a hot-water immersion treatment without the RF heating. For example, a treatment offered for apples (mean weight 266 g) consisted of heating the fruit in a hot water bath at 45°C for 30 min, transfer to a 45°C water tank in an RF heater to raise the water temperature via RF heating to 48°C and maintaining the water and apples in this 48°C water for an additional 15 min (Wang *et al.*, 2006). This complex treatment provided complete kill of codling moth, *C. pomonella*, in small-scale tests. For large-scale confirmatory testing required to substantiate a phytosanitary treatment the severity of the treatment would most likely need to be increased to be successful. This RF-assisted hot water treatment might be no better in time or surely in simplicity and cost than a simple hot water treatment, which was not compared in the research. It would behove researchers to compare novel treatments for efficacy, ease and cost with a reasonable standard, whether that standard is commercially used or not.

A similar but more favourable development occurred with RF heating of walnuts to control a variety of stored-product pests (Mitcham *et al.*, 2004; Wang *et al.*, 2006). Like RF heating of fruits, it may be done most reliably by combining another source of heat, in this case heated air, with RF heat. It remains to be seen, however, if this combination treatment would be more favourable for walnuts than a simple heated-air treatment.

Recent studies have also focused on RF heating of wood to destroy wood-infesting insects, such as drywood termites (Lewis *et al.*, 2000). Death of Asian longhorned beetle, *Anoplophora glabripennis*, larvae and pupae occurred after 3 min in 'green' poplar blocks and only 5 s in dry blocks heated in a 2.45 GHz machine, demonstrating in this case at least a clear advantage to low moisture for achieving quick kill (Fleming *et al.*, 2003). RF heating is being considered as a phytosanitary treatment for use against quarantine pests of wood packaging materials (IPPC, 2007).

Ohmic heating

Ohmic heating works by passing an electrical current through a resisting substance, which causes the substance to heat. This is the technique employed by electric heaters. In the case of ohmic heating of food, the food serves as the resistor, and therefore, the heating element. Ohmic heating is the most efficient way to convert electrical energy to heat. It is used commercially for microbial inactivation in some foods, such as whole and sliced fruits and liquid egg and causes less damage to the foods than conventional heating methods (Ruan *et al.*, 2004).

Because killing insects requires less heat than inactivating microbes, ohmic heating might be a viable phytosanitary treatment if it can be shown not to harm fresh commodities. Ohmic heating of particulate matter, such as whole fruits, must occur in a liquid medium. Therefore, to heat whole fruits and vegetables via ohmic heating would require treatment in water. Preliminary research by one author of this book (Hallman) has shown that ohmic heating can function more cheaply as a phytosanitary treatment for fresh commodities than radiofrequency heating. They both suffer from heating variability. By the same token, if radiofrequency heating of fresh commodities deserves further research, then ohmic heating certainly does also.

Thermometry and Heat Regulation

Equipment is currently available to register temperature in laboratory experiments as precisely as 0.01°C increments and to regulate heat to within ± 0.1°C provided that the equipment is kept properly calibrated. Temperature regulation at this level of precision in commercial operations could prove impractical and a more useable precision level might be 0.5°C. Phytosanitary requirements may specify a mean temperature value with an allowable ± variance but are more likely to specify a minimum temperature, in which case the temperature variance needs to be added to the minimum figure setting. For individual commodity units surface and/or core temperatures may need to be measured depending on the target organism and the commodity. For example, heated-air treatments usually rely on centre or seed surface temperatures to determine when the treatment is finished.

Surface temperatures may be required to determine how much humidity to add to the system to prevent the commodity from losing moisture or moisture from condensing on the product. When temperature probes based on platinum resistance temperature detectors (Fig. 8.4), thermocouples or other equally precise technologies linked with microprocessors to control heating and prevent over-run, the resultant fast reaction enables excellent temperature control.

Mode of Action on Arthropods

Lethal heat stress in insects and other arthropods results from time at an injurious temperature (i.e. it is a time × temperature effect). Heat affects neurosecretory processes, modifies the constitution of proteins and can stimulate the development of heat shock proteins that influence the response of the organism. Raised temperatures can cause a reduction in haemolymph pH in arthropods with resulting changes in levels of minerals, free amino acids and blood sugars (Denlinger and Yocum, 1998). At the macromolecular level, heat is known to cause quantitative and qualitative changes in protein production, producing a set of heat-induced proteins including those known as heat shock proteins, which may impart increased tolerance to heat as well as other forms of stress (Lurie and Jang, 2007). Heat may also cause lesions in DNA and unfolding and hence inactivation of proteins.

A couple of models propose that the weak link leading to cell death is heat-induced disruption of the plasma membrane that sets in motion a cascade of events involving inactivation of membrane proteins, leakage of ions in and out of cells and disruption of bioelectrical processes, leading to inappropriate activation of enzymes, further breakdown of cell function and structure leading to death (Denlinger and Yocum, 1998). The organism's death then occurs when enough cells from a critical, heat-susceptible biological process are destroyed. Heat also promotes rapid desiccation in small organisms; however, this probably does not contribute to heat-induced mortality in phytosanitary treatments because the pests are usually maintained in moist conditions throughout the treatment.

Lethality can be acute or chronic. However, the end point is usually acute mortality of any stages present as inspectors do not generally accept the presence of live quarantine pests after a heat treatment upon inspection.

Heat phytosanitary treatments are usually limited more by the thermotolerance of the commodity to be disinfested. Although any temperature above the threshold for pest survival and below the acceptable injury threshold for the commodity could be used, commercial considerations normally favour the most rapid treatment.

Although some reports that eggs and early instars may be more tolerant of heat in vitro than last (third) instars fruit flies, third instars are usually treated as the most heat tolerant during in situ phytosanitary research because they may be found deepest in a fruit, hence, will be exposed to less total heat stress than stages closer to the surface of the fruit. Furthermore, third instars are closest to reaching the reproductive stage, having less developmental hurdles left to surmount than eggs and early instars. For treatments that may heat the commodity differently (not always from the outside in), such as radiofrequency heating, third instars should not necessarily be considered the most tolerant.

When it comes to determining the most tolerant stage of a pest to a phytosanitary treatment, researchers should choose reasonable end points. The most reasonable end point may be different for different stages. For example Jang *et al.* (1999) used larval movement after 24 h as the measure of survival for first instar tephritids subject to heat and puparial development as the measure of survival for third instars. The first instar had fewer hurdles to surmount and was deemed the most thermotolerant of the two stages in three of four species studied. Choice of end point for survival depends on the objective of the research. For phytosanitary purposes it should be required that first instars achieve a greater level of development to be counted as survivors, one that would represent a risk more akin to pupariation for heat-treated third instars.

These studies enabled the comparison of responses of species, developmental stages and ages within stages and their relationship to developmental parameters, showing that early in development within a stage, susceptibility to heat is greater than subsequently. Nevertheless authorities of most countries require simply identification of the most susceptible stage and its use to determine the efficacy of any proposed treatment.

Differences in susceptibility to heat treatments can be expected between species. Difficulties arise in making comparisons based on results of more than one author frequently due to differing lethal time (LT) values used in reporting results. Sometimes *C. capitata* is shown to be less susceptible than the major pest species, *Bactrocera tryoni*, *B. dorsalis* and *B. cucurbitae*, although these differences may not be statistically significant (Armstrong *et al.*, 1989). However, the differences are not necessarily genera related, with the Australian species *Bactrocera jarvisi* found by Corcoran (2002) in vitro to be markedly more resistant to heat than *B. tryoni*, and *Bactrocera cucumis* much less so. This study, using very precise temperature and timing control, found considerable difference in heat tolerance of eggs and larvae at differing developmental ages within each stage. For a given treatment on eggs of *B. tryoni* mortality was only 10% in eggs which were 60% developed compared to 80% at 80% egg development. Of course, variations in the way research was conducted by different researchers can be responsible for a large portion of any differences observed among species.

Heat treatment of commodities has two definable timed stages, the approach time and the holding time. Each time contributes to the required ultimate lethal response of the pest. But post-treatment time until temperatures drop below lethal levels may also contribute to mortality. Hydrocooling of hot-water treated mangoes must not be done until 30 min after the treatment is terminated; if it is done earlier 10 min must be added to treatment times (APHIS, 2007).

Apart from phytosanitary studies, most other heat tolerance studies reported (such as Chapman, 1998) are from physiological and ecological studies and relate to survival, not intentional mortality. This research may be useful to heat phytosanitation. Responses of tephritid fruit flies to heat at disinfestation temperatures have been modelled (Tang *et al.*, 2007). Models can enable prediction of the temperature range applicable to large-scale experiments where these are required and accurately identify commonality of lethal responses across related species and more widely, across the range of pest species which might be associated with a commodity.

Heat acclimation is a phenomenon that can influence tolerance of a pest to heat. It is a physiological response to stress resulting in the production of protective heat shock proteins. In at least some instances these can be identical to stress proteins developed as a result of cold stress. It may not occur in all species but nevertheless has been reported from single-celled organisms such as yeasts, as well as insects (Parsell *et al.*, 1993; Lester and Greenwood, 1997; Lurie and Jang, 2007). Heat acclimation is more likely to occur where the approach time to a heat treatment is relatively long, as heat shock synthesis requires preconditioning time at a sublethal temperature. The possibility of heat acclimation should be taken into account in the technology transfer from experimental to commercial operation where heating patterns differ.

As with virtually all other treatments except pesticides, there is no residual protection with heat. Rigorous control of process is required to ensure that untreated product cannot be accessed for packaging before treatment (i.e. cross-contamination of the product flow must be prevented by a fail-safe system). It further requires that the product must be packaged in a facility protected from external pest contamination and that packages must exclude re-infestation during transit if the pest has that capability.

Commodity Tolerance

Commodity tolerance is of primary importance for fresh commodities and propagative materials; they are alive and actively metabolizing just as are the pests infesting them. Commodity tolerance is often the principal limiting factor on the use of heat treatments of fresh commodities. Of course, inanimate commodities may be damaged by heat, also.

Commodity tolerance is not of concern to regulatory agencies although they make an effort to provide phytosanitary treatments that provide a commodity of acceptable quality. Regulatory agencies should not be held liable for damage caused to commodities by treatments that they accept. It is incumbent on those with economic interests in the commodity being treated to be responsible for the quality of commodities treated on a commercial scale, which might differ from those treated on a small, experimental scale. Pilot commercial trials to judge commodity quality under commercial conditions of export and storage should be conducted.

Tolerance of fresh horticultural commodities to heat treatment is influenced by a variety of variables including developmental stage. In general for heat treatments, the riper the fruit at treatment the lower is the risk of heat injury. However, heat treatments are often done on unripe fruit that ripens in the marketing channel. Heat treatments may accelerate ripening and reduce shelf life. Heat treating fruit at the earliest possible stage of maturity is often done to maximize the time available to get it to the consumer. In general, temperate pome and stone fruits do not tolerate heat as well as tropical fruits. Heat may accentuate the appearance of bruises and other blemishes. In a study by Jacobi *et al.* (1994) production region and cultivar were found to influence the incidence of heat treatment injury in mangoes.

The three criteria that are most important in evaluating treated edibles are gross appearance, organoleptic qualities and shelf life. Incredibly, shelf life is not noted for most commodity quality studies, and organoleptic evaluation is often lacking. These tests may be done without scientific equipment by heating lots of at least 20 individual pieces with treatment conditions approximating to commercial heating rates, storage temperatures and times. Studies should be replicated across the bulk of commercial commodity variables including cultivar, growing area, season, size or any factor that may affect quality. It would be ideal if those doing the evaluations did not know which lots were treated and which were controls and that the evaluators are individuals whose abilities include judging the quality of commodities, such as produce brokers, importers, retail managers and informed consumers. They would judge the commodity based on criteria that they would use in deciding to buy or consume it: appearance, both exterior and interior, firmness, texture, both in the hand and in the mouth, smell and flavour. Some detectable differences may be inconsequential. Shelf life would be evaluated by holding a part of the lot at typical storage conditions for that commodity, be it in a storage facility, retail store or home, until the commodity is no longer useful.

Experimental measurements such as Brix, pH, colour, firmness, citric acid content and electrolyte leakage are unnecessary unless comparisons or conditions that require such measurements are being investigated: for example, if a consumer group wishes to know whether a heat treatment affects vitamin content or if a researcher is investigating why a fruit loses considerable shelf life after treatment. Non-edible commodities would be evaluated similarly, except for taste, of course. The emphasis with cut flowers may be on appearance, smell and certainly, shelf life.

Even if results of quality trials have been reported by researchers, it would behove industry to do some type of quality trials themselves to be sure that they accept any effect the treatment has on the commodity. Industry's criteria for acceptance may be different from the researchers'.

Conclusions

Disinfestation of food commodities with heat to satisfy phytosanitary requirements has the advantage of freedom from chemical residues and satisfies consumers generally concerned in this way as well as those who are committed to 'organically produced' foods. It has the potential to disinfest commodities to the highest standards of quarantine security but requires a temperature-time 'window of opportunity' in which a pest is killed but where there is no unacceptable treatment injury to the commodity. Treatment time is typically short and capital cost can be low if simple hot water dips are satisfactory. Disadvantages include the need either to treat the produce immediately after harvest or it must be warmed and re-cooled. Hot air treatment of fresh commodities is complex and relies heavily on modern electronic technology. Consequently it involves expensive standing equipment, although less complex than irradiation with a gamma- or X-ray source. We see heat as an essential option for disinfestation for phytosanitary purposes into the foreseeable future.

9 Phytosanitation with Ionizing Radiation

Ionizing radiation is caused by particles or that portion of the electromagnetic spectrum that is high enough in energy to break chemical bonds. Ionizing particles include electrons, protons, other ions, neutrons and photons. The ionizing portion of the electromagnetic spectrum is visible light and shorter wavelengths. However, only certain chemicals, such as those involved in photosynthesis and those used for photography, are ionized by visible light. Ultraviolet light ionizes, being responsible for the skin damage of sunburn and some skin cancers, and is used to disinfect food and water and to sterilize work areas. It has also been researched as a surface phytosanitary treatment.

The sources of ionizing radiation approved for food and, thus, phytosanitary purposes, are ultraviolet, electron beam, bremsstrahlung X-ray and gamma rays from the ions cobalt-60 and caesium-137. Very high energy electron beams and X-rays can make an irradiated product radioactive, so the energy level is restricted to a maximum of 10 MeV for an electron beam and recently, for X-ray, in the USA was raised from 5 to 7.5 MeV if tantalum or gold targets were used (FDA, 2004). Ions emit the same energy levels independent of the quantity of material present. Cobalt-60 has a mean gamma energy of 1.25 MeV, and for caesium-137 it is 0.66 MeV. With these energy levels, neither can cause a treated product to become radioactive.

Penetration by electron beam is much shallower than gamma rays or X-rays and can only be used to treat produce loads of no more than 5–10 cm in depth, such as single layers of fruit, berries in shallow clam-shell containers, a thin stream of grain or shallow boxes of cut flowers on a conveyor. Low energy electrons (≤ 0.3 MeV) have low penetration, although they will surface-treat commodities exposed on all sides. Hayashi *et al.* (2004) could not control larvae of stored-product pests inside grain and seeds using soft electrons at 60 keV.

Food irradiation causes some concern among portions of the public and some food marketers, but there is no scientific support for the contention that the food irradiation process at any dose may be unhealthy (WHO, 1999). Furthermore, it is nigh impossible for machine sources (electron beam and X-ray; Fig. 9.1), built to

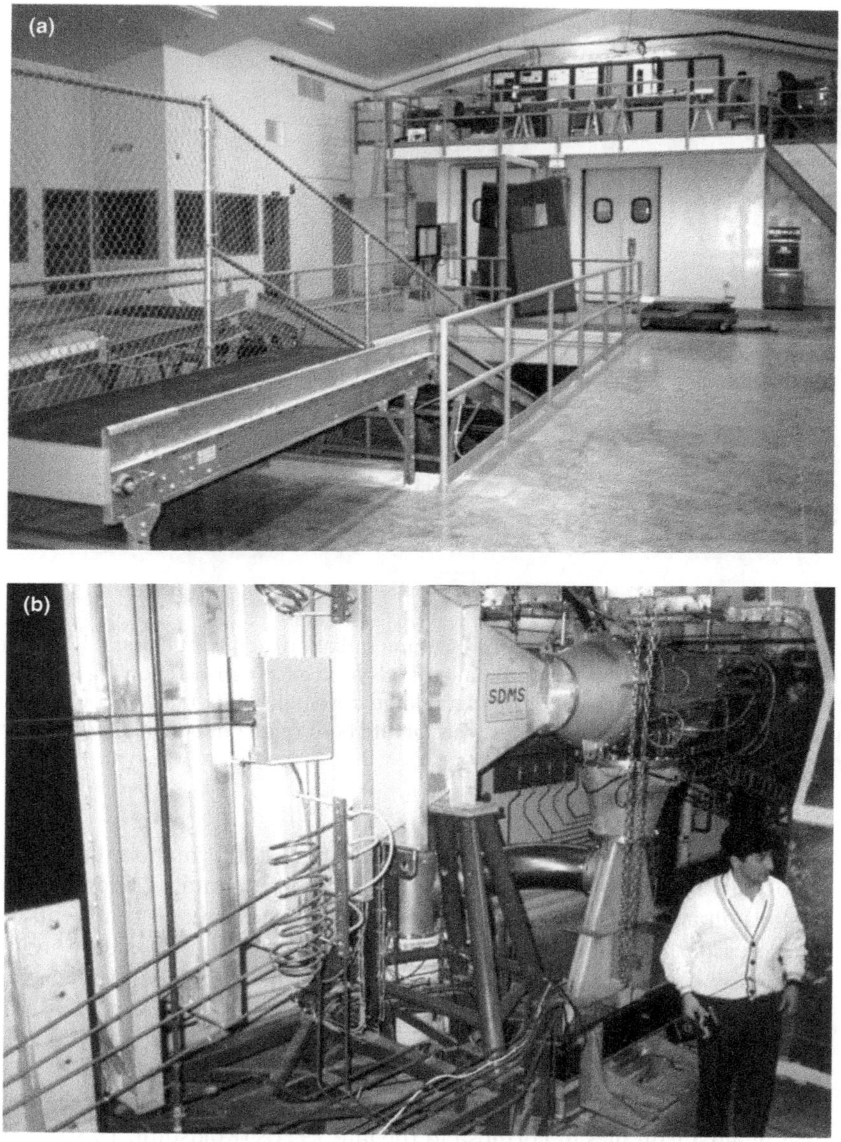

Fig. 9.1. An X-ray irradiator facility and electron accelerator unit, Florida, USA. (a) Entry and exit are divided by a barrier to keep treated and untreated product separate. (b) Electron accelerator.

deliver specific energy levels, to malfunction in a way that provides dangerously high energy levels. Also, there is no functional way for cobalt-60 or caesium-137 ions in food irradiation facilities (Fig. 9.2) to contaminate food being irradiated, barring a major catastrophic event and that would render the facility inoperable.

The International Database on Insect Disinfestation and Sterilization (IDIDAS) contains more than 3300 references to technical irradiation studies on more than

Fig. 9.2. A multi-purpose cobalt-60-sourced gamma radiation facility in Florida, USA, showing carriers in the loading area. Two standard pallet-loads fit in each carrier, allowing for economical treatments, but possibly resulting in large dose uniformity ratios (about 2.5) even though product is irradiated on both sides. This facility has done commercial phytosanitary treatments of fresh produce since 1999.

300 species of arthropods with further links to pest information through EcoPort (EcoPort, 2007; FAO/IAEA, 2007). This database contains references to both phytosanitary treatments against pests and sterilization treatments for the purposes of pest management using inundative release of sterile individuals, known as the sterile insect technique (SIT) or the sterile insect release method (SIRM). Because quarantine security can be satisfied by reproductive sterilization and this may be indicated through phytosanitary research, there is relevance between the two groups of references.

The first notion that irradiation could be used as a phytosanitary treatment was against fruit flies in Formosa in the late 1920s (Koidsumi, 1930). Two basic tenets of irradiation as a phytosanitary treatment were begun at that time. Acute mortality was not necessary to provide quarantine security and prevention of adult emergence could be accomplished with much lower doses. The fruit flies studied increased in radiotolerance as they developed, thus, third instars would be the most radiotolerant stage found in fruit.

The phytosanitary world pursued other treatments, such as cold, vapour heat and fumigants, and 25 years would pass before irradiation was studied to a significant extent again. However, irradiation was not tried commercially until 1986, 56 years after Koidsumi's (1930) observation, when one load of mangoes was irradiated in Puerto Rico and shipped to Florida, USA, for sale. This was in response to the loss of ethylene dibromide as a phytosanitary fumigant for fruit flies in mangoes. Although the response by consumers was reported to be positive (Phillips, 1986), no further shipments were made because of the lack of a regulatory framework in dealing with any live insects found after irradiation and the development of another alternative to ethylene dibromide for mangoes, hot water immersion (Chapter 8). After a single shipment of Hawaiian papayas to the US mainland for irradiation in 1987 and the posting of an irradiation protocol for papayas in 1989 that was never used, no more fruits were shipped using phytosanitary irradiation until 1995. The European and Mediterranean Plant Protection Organization (EPPO) approved irradiation doses for several species of arthropods on cut flowers in 1993 based on results from one study, although to our knowledge that treatment also was never used (EPPO, 2007b). These are the only phytosanitary irradiation doses that EPPO has set. Some of the doses are open to question if they are to be used because of the small numbers of test insects and the fact that no margin of error was added to the doses to compensate for insufficient data.

A shipment of 240 boxes of Hawaiian papayas to a cobalt-60 irradiation facility near Chicago on 5 April 1995 marked a new era in phytosanitary irradiation (Hallman, 2001b; Moy and Wong, 2002). Amount, pest species controlled, species of commodities treated and sales outlets of Hawaiian fruit shipped to two cobalt-60 irradiation facilities expanded over the next 5 years until an X-ray facility was opened in Keaau, Hawaii in August 2000.

The X-ray facility was constructed in part with a US$6.75 million loan from the US Department of Agriculture (USDA), Rural Development Business and Industry Guaranteed Loan Program to support commercial use of the technology. Capacity of the X-ray facility is 13,000 t/year. At one time about 50% of papayas leaving Hawaii were irradiated, but that number is down to 10–15% because occasional extended plant shutdowns for maintenance over the years have caused some marketers to utilize vapour heat treatment for more fruit (Anon., 2006b).

Although the state of Hawaii is noted as a site of key motivation in commercialization of irradiation as well as other phytosanitary treatments, accomplishments in the state of Florida concerning phytosanitary irradiation are generally overlooked. Florida was the first recipient of commercial fruit irradiated for phytosanitary purposes (mangoes from Puerto Rico in 1986 against tephritid fruit flies). The first commercial irradiation facility built expressly for phytosanitary purposes was in Florida. It was completed in 1992 with the idea that irradiation would replace ethylene dibromide fumigation as a phytosanitary treatment for grapefruit although in practice other alternatives were used. That facility currently performs irradiation on a number of products, including for phytosanitation, and may be the commercial facility with most diverse uses of food irradiation in the world.

The first commercial use of phytosanitary irradiation against a pest that occurs on shipped commodities in the adult stage (sweetpotato weevil, *Cylas formicarius elegantulus*) was in Florida. This was a major step for a treatment that leaves pests alive for a longer time after treatment than any other commercially used phytosanitary treatment (Hallman, 2001a).

Florida has the potential to utilize this treatment to a greater extent. It has a dedicated, large-scale cobalt-60 irradiation facility (Fig. 9.2) that can treat a larger quantity of products more economically than competing X-ray facilities. A broad range of commodities needing phytosanitary treatments, and irradiation may be the only one that works for some of these commodities, is grown in the state. Where Caribbean fruit fly, *Anastrepha suspensa*, is the only quarantine pest present, Florida has phytosanitary approval to ship any irradiated fruit to California although only guavas are irradiated on a continuing basis.

Phytosanitary irradiation research has been summarized in recent reviews. Follett and Griffin (2006) discussed treatment development methodology and gave a detailed history on the international regulatory progress of phytosanitary irradiation. Hallman (2006) presented an explanation of the international standard, ISPM 18 *Guidelines for the Use of Irradiation as a Phytosanitary Measure* (IPPC, 2007). Bakri *et al.* (2005) summarized the aforementioned online database (IDIDAS) that strives to collect all publications dealing with phytosanitary irradiation. Hallman (1999) and Hallman and Loaharanu (2002) reviewed the large body of literature on tephritid fruit flies and concluded that a generic, default dose of 150 Gy as proposed by the former International Consultative Group on Food Irradiation (ICGFI, 1994) is justified. Hallman (2001b) contrasted irradiation with other phytosanitary treatments, gave a history of the treatment until 2001, and identified important aspects of irradiation phytosanitary treatment research and future research needs. Hallman (2000) reviewed the literature on irradiation of pests other than Tephritidae and presented conclusions that would facilitate phytosanitary irradiation research. These included: the most tolerant insect stage to irradiation is that which is most developed; female insects are reproductively sterilized with doses equal or less than males; and insects in diapause are not more radiotolerant than the same species not in diapause.

Irradiation will be used most often as a single treatment although a combined treatment such as with cold storage or heat may be advantageous where there is product intolerance. Irradiation causes only minor change in the temperature of the treated commodity so it can be done after cooling to normal handling temperatures. Gamma- or X-ray irradiation can be done while the commodity is in merchantable packaging which has post-treatment security advantages.

Irradiation is different from all other commercially used treatments in one major way: it is the only technology that does not cause acute mortality at doses which can confer quarantine security. Irradiation doses required to kill every pest present in less than a day are usually too high for most produce to tolerate without unacceptable effects on commercial quality (Fig. 9.3). However, much lower doses (50–350 Gy) can cause total mortality after some time or result in total sterility, which is equivalent to mortality in preventing the next generation. In practice, the aim of an irradiation quarantine treatment is usually to cause mortality of immature stages before development to the adult or, where adults or pupae are present, to stop or prevent reproduction. Usually, the dose required to prevent late pupae from emerging as adults is prohibitively high relative to the tolerance of most fresh commodities (Hallman and Hellmich, 2007). Therefore, prevention of reproduction by emerging adults is the most reasonable objective of

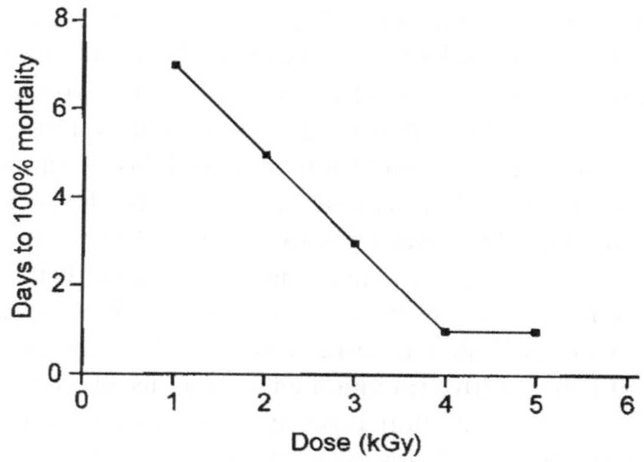

Fig. 9.3. Days to reach 100% mortality of Indian meal moth, *Plodia interpunctella*, larvae after irradiation (Source: after Saeed *et al.*, 2006). At 0.25 and 0.5 kGy, respectively, mortality reached about 35 and 90% after 21 days.

an irradiation treatment against insects that pupate in the commodity, such as moths of the family Pyralidae.

A disadvantage of a treatment that allows for live, but reproductively sterile, adults is the risk that some of these adults might be found in survey traps, which could trigger costly regulatory responses. But allowing for the presence of live adults after treatment is necessary for irradiation to be realistically used for pests that may be present as adults or late pupae in shipped commodities, which are most pests. However, these adults may be less hardy, with shorter lifespans, and less likely to find their way into survey traps. Furthermore, upon examination it may be determined that they have been radio-sterilized and pose no threat (Dyck *et al.*, 2005).

Irradiation is the most recent commercial phytosanitary treatment for fresh commodities. The first continuous application did not begin until 1995. It has been approached with caution not only because of perceived objections to food irradiation by marketers and the public (Eustice and Bruhn, 2006), but also by the regulatory community because it was considered to be the only commercially applied treatment that did not result in almost immediate mortality of quarantine pests, thus inspectors could expect to find live pests upon inspection even days after treatment. Therefore irradiation is being developed with the latest understanding of phytosanitation and, consequently, more caution than possibly was applied to previous treatments. As such it could serve as a model for regulation and research with other treatments. *Guidelines for the Use of Irradiation as a Phytosanitary Measure* (IPPC, 2007) is the first phytosanitary treatment standard promulgated by the International Plant Protection Convention (IPPC) and was organized in a way that could serve as a template for future standards (Hallman, 2006). Irradiation treatments are also the first treatments being considered for a new IPPC international phytosanitary treatment manual.

Irradiation phytosanitary treatments should be considered with more care than other treatments not because the treatment itself presents some health or environmental risk, such as led to restrictions on the use of the fumigants ethylene dibromide and methyl bromide (see Chapter 10), but because of the risk associated with the lack of an independent verification of efficacy all other treatments possess. All other commercial treatments except irradiation are expected to have resulted in essentially 100% mortality when treated commodities arrive at inspection ports. Most pests irradiated with the minimum absorbed dose for quarantine security will be alive for some days after irradiation. Therefore, inspectors cannot assume treatment failure when live pests are found in irradiated commodities. The discovery of live pests following cold, heat and fumigation treatments has essentially been the only way that treatment failure was identified. If it is generally the case that discovery of live insects after treatment is the only measure of treatment failure then treatment failure with irradiation will not be discovered.

Approval of an irradiation dose against mango seed weevil, *Sternochetus mangiferae*, is illustrative of this concern. In 2000, 100 Gy was proposed as a minimum phytosanitary dose against this insect because research 'demonstrated that the weevils are effectively killed or sterilized at this dose' (APHIS, 2000b). After a public comment period the dose was finalized in 2002 at a higher level, 300 Gy, because, 'the only research that found 100 gray to be effective ... was a limited study involving a very few insects' (APHIS, 2002b). The final rule went on to state, 'other research ... found that a dose in the 300 gray range was necessary to effectively control the weevil'.

This treatment was designed to allow commercial irradiation and shipment of mangoes from Hawaii to the US mainland. But its publication could lead to US importation of mango seed weevil-infested mangoes from other parts of the world as well. It has not been used commercially because mangoes from Hawaii cannot compete economically with those available to the mainland USA from other countries. It is a relatively safe example of the need for approaching irradiation phytosanitary treatments with more care than other treatments, in that the insect is probably not a great threat to US agriculture because it is specific to mangoes (Follett, 2001), so if the treatment were used at 100 Gy and failed to completely prevent reproduction of the pest, the damage would not be great.

The mango seed weevil does not normally damage mango pulp in any case (Follett and Gabbard, 2000). But, this is a good example to show that research on irradiation phytosanitary treatments against pests must be done with a greater degree of certainty than acceptable for other treatments where almost all insects can be expected to be dead upon inspection. What approval of an irradiation dose for mango seed weevil based on 'limited study' (APHIS, 2002a) did accomplish, perhaps unwittingly, was to set a precedent. There are many other pests for which as much or more research has been conducted, and it can be argued that doses for these could also be set.

Lack of acute mortality is associated with another disadvantage of irradiation compared with other treatments. Contaminating or 'hitchhiker' pests are sometimes found in commodities treated for known quarantine pests. If the contaminating pests are dead no further regulatory action may result. With irradiation they will most likely not be dead, and barring reliable data showing that the dose applied controls the contaminating pests the consignment will probably not be released without re-treatment.

Dosimetry

Dosimetry is the basis for assurance that the qualification of the source and the treatment process have been done in accordance with phytosanitary requirements. Depending on the level of treatment applied differing criteria need to be selected for estimation of quarantine security. These can range from prevention of eclosion of eggs through to prevention of development beyond a particular pre-adult stage or prevention of reproduction by adults. The problems which can arise when live individuals are found during phytosanitary surveillance sampling require the provision of reliable assurance of treatment at a level in excess of the minimum for quarantine security and of prevention of infestation subsequent to treatment.

The Système International (SI) unit for absorbed dose of irradiation is the gray (Gy), named after the British radiobiologist Louis Harold Gray in 1975. One gray is 1 J of energy absorbed by 1 kg of material. Earlier literature used the rad (radiation absorbed dose) as the standard unit: 1 rad = 0.01 Gy. Literature using rad is still occasionally published. The röntgen (R) was used in some early literature (1 R ≈ 9.33 mGy) but its use to represent absorbed dose is discouraged.

Absorbed dose can be measured using a number of technologies and is the subject of international standards (IAEA, 2001; ISO, 2002). Methodologies vary in precision. The absolute methods for measuring absorbed radiation dose are conducted at national standards laboratories through calorimetric determination of heat produced or by the measurement of the number of ions produced in a gas under standard conditions. These absolute dose measurements are transferred to radiation processing applications through reference class dosimetry systems. For reasons of practicality, secondary or routine systems of dosimetry are used in commercial radiation processing applications.

Dosimeters, regardless of the class or composition, are devices that when irradiated, exhibit a quantifiable change in some property which can be related to absorbed dose using appropriate analytical instrumentation and techniques. Many routine dosimetry systems are available for use in the dose range of interest (50–300 Gy) for phytosanitary treatments. Examples include the Fricke ferrous sulphate and the ceric/cerous dosimeters, which are chemical-based liquid dosimeters; polymethylmethacrylate (PMMA) and radiochromic optical waveguide dosimeters, which are chemical dye polymer-based dosimeters; and alanine dosimeters, which are amino acid-based dosimeters (Fricke and Hart, 1967). With the exception of alanine dosimeters, all systems are based on a change in optical density following an irradiation event. This change is measured using a visual spectrophotometer or photometer at a specific wavelength. Alanine dosimeters are assayed using electron paramagnetic resonance (EPR) spectroscopy to determine the level of free radicals derived from the alanine following an irradiation event. All routine systems, when calibrated and maintained in accordance with international standards, offer an acceptable precision for most applications. Label dose indicators or radiation-sensitive indicators, that change colour upon exposure to a minimum dose, and clear PMMA dosimeters, currently lack the precision required for confirmation of phytosanitary treatments. Another method, the induced transient electric current across the pn junction of a bipolar power transistor (diode) in plastic packaging is useable in the range of interest. Promising new dosimetry systems are emerging including Sunna films, which have the advantage of fluorometric read-out (IAEA, 2001).

Radiation is attenuated at a rate dependent on the density of the commodity being treated, resulting in varying absorbed doses throughout the treatment unit. This difference in dose is described as the dose uniformity ratio (DUR). The pattern is nonlinear and its geometry can be determined only from single point determinations on a regression line. So the maximum level of the treatment dose found to be effective in experiments against a pest should become the minimum used operationally in commercial practice. Where an experimental sample is small and dosimetry is accurate, the DUR can be as low as 1:1.1. Commercial irradiators, which can handle much larger volumes, may operate with DUR as high as 3:1, even where multiple passes by the source are made on each side of the commodity treatment unit.

Mode of Action on Arthropods

As a broad principle, the tolerance of organisms to irradiation increases with development. Hallman (2000) reviewed doses required to prevent development and/or reproduction among a wide variety of arthropods and found only one instance, human body louse, *Pediculus humanus humanus*, where the literature indicated that an earlier stage might be significantly more radiotolerant than a subsequent stage when the same measure of efficacy was used.

Mature somatic cells are less sensitive to radiation than mitotically active stem cells. An acute lethal dose involves somatic cell systems as well as stem cells but requires higher dose levels than an outcome of reproductive sterilization. Dose levels that are not acutely lethal can be lethal eventually where they act on cell systems that maintain mitotic activity. Mitotically active reproductive cells are the most radiosensitive although they have differing killing times or sterilization susceptibilities according to the developmental stage subjected to irradiation. This characteristic is more important to sterile insect technique (SIT) programmes where the pest is irradiated at a pre-adult stage then develops to the adult which is required to be biologically competitive with wild non-irradiated individuals. Such studies provide a measure of the minimum dose that will achieve quarantine security through sterilization. Mitotic activity levels in gonads of sexual species are almost certainly the reason for the differing sterilization dose thresholds between males and females, with females in general the more radiosensitive (Hallman, 2000; Bakri *et al.*, 2005).

The most radiosusceptible cells in pre-adult stages appear to be those that become mitotically active at each moult including those associated with neurosecretory hormone production. This can result in death of irradiated pre-adults at the next or a subsequent moult or change of stage (e.g. eclosion, moult, pupation or emergence of the adult). Damage to somatic cell tissues in pre-adults is often not manifested until the adult stage when abnormalities become apparent. Abnormalities such as loss of normal flight ability may preclude reproduction, thus conferring quarantine security even where gonial sterility is not complete. Regenerative cells for the midgut lining remain mitotically active during the life of the adult and are a site highly susceptible to radiation. For species such as the flour beetle, *Tribolium confusum*, adults of which have a normal lifespan of up to 1 year, failure of nutrient uptake will cause early death within a few weeks or less (Szczepanik and Ignatowicz, 1994). However, for species that do not feed as the adult, such as some moths, this effect would not be significant. In a few cases irradiated adults lived up to 29% longer than non-irradiated controls (Hallman, 2000). It may be that removing the strain of reproduction increases longevity in certain organisms.

Physiological changes which might serve as markers indicating treatment with irradiation have been studied in response to concerns that live insects found at inspection of commodities required to be irradiated could indicate omission of treatment (Nation *et al.*, 1995). Nation *et al.* (1999) nominated seven parameters including whole body melanization, phenoloxidase assays, supraoesophageal ganglion/proventriculus ratio, imaginal disc ratio, haemocyte count and larval weight as possible indicators of irradiation in juveniles or adults of tephritid fruit flies, moths and beetles. Other possible indicators are body pigmentation such as humeral callus colour in fruit flies (Lin and Heather, unpublished research) and changes in midgut histology of beetles and moth larvae (Ignatowicz, 1999). Rahman *et al.* (1990) found that in larvae of Mediterranean fruit fly, *Ceratitis capitata*, irradiated as young larvae, development of the supraoesophageal ganglion was subsequently inhibited and that this could be determined from the size of the ganglion relative to the unaffected proventriculus. Subsequently this effect was found in species of *Bactrocera* (Rahman *et al.*, 1992; Lescano *et al.*, 1994).

Unfortunately these indicators require a certain amount of time at temperatures allowing further development for them to become manifest. Thus, they are not useful at inspection stations that must make quick decisions on treatment efficacy when there has been insufficient time for indicator development since treatment or if commodities have been stored at cool temperatures after irradiation. Expression of these physiological markers is dose, age and stage related and none can give more than a positive indication when they are expressed. Also, they may not be as well or as consistently expressed in groups other than tephritid fruit flies (Heather *et al.*, 1999; Ignatowicz, 1999).

Furthermore, the best that a physiological or any other indicator, including proper dosimetry, can do is to indicate that the subject was irradiated. Alone, it does not address any other concern, such as whether the treatment as applied was adequate, or that some other factor not considered during the research or that may have changed since the research was conducted renders the treatment sub-efficacious. For other phytosanitary measures (heat, cold, fumigation) the presence of dead insects found upon inspection in treated commodities is virtually the only independent measure indicating that the treatment was efficacious.

The best way to manage the chronic mortality characteristic of irradiation treatment could be through reliably certified treatment, using a calibrated dosimetry system coupled with pest management during production and handling, which ensures that as far as possible the incidence of the pest is below the level of detection at any sampling by regulatory authorities. It should be a tenet of phytosanitary management that quarantine treatments are not a replacement for operational pest management nor should they be considered to be stand-alone procedures. Postharvest treatments must always be part of a larger system that ensures quarantine security including production, harvest, packing-shed management, culling, cleaning, packing, post-packing management and travel to market.

Irradiation Doses for Quarantine Pests

Phytosanitary irradiation research has been done with a variety of quarantine pest groups, especially tephritid fruit flies. Control doses for a number of horticultural species other than fruit flies are given in Table 9.1 and stored-product pests in Table 9.2.

Table 9.1. Phytosanitary irradiation treatments for some major horticultural pests other than fruit flies. Refer to IDIDAS (FAO/IAEA, 2007) for detailed information on dosages and comparisons between related species, within genera and within families.

Order	Pest species	Host commodity	Effective dose (Gy)	Reference
Lepidoptera	*Cryptophlebia ombrodelta*, Macadamia nut borer	Lychee, macadamia nut	250	Follett and Lower (2000)
	Cryptophlebia illepida, Koa seedworm	Lychee, macadamia nut	250	Follett and Lower (2000)
	Cydia pomonella, Codling moth	Pome fruits	200	Mansour (2003)
	Epiphyas postvittana, Lightbrown apple moth	Pome fruits	199	Dentener *et al.* (1990)
	Grapholita molesta, Oriental fruit moth	Many fruits	200	Hallman (2004a)
Coleoptera	*Asynonchus cervinus*, Fuller rose beetle	Citrus	150	Johnson *et al.* (1990)
	Sternochetus mangiferae, Mango seed weevil	Mango	300	Follett (2001)
	Cylas formicarius elegantulus, Sweetpotato weevil	Sweet potato	140	Follett (2006a)
	Conotrachelus nenuphar, Plum curculio	Stone fruits	92	Hallman (2003)
Hemiptera, Homoptera	*Brachycorynella asparagi*, Asparagus aphid	Asparagus	100	Halfhill (1988)
Thysanoptera	*Thrips palmi*, Melon thrips	Fruits, cut flowers	350	Bansiddhi *et al.* (2004)
Acarina	*Brevipalpus chilensis*, False red mite	Grapes	300	Castro *et al.* (2004)
	Tetranychus urticae, Two spotted mite	Flowers, fruits, vegetables	300	Goodwin and Welham (1990)

Table 9.2. Comparative phytosanitary irradiation treatments for some pests of stored products and other durable agricultural commodities. Refer to IDIDAS (FAO/IAEA, 2007) for detailed information on dosage and comparisons between related species, within genera and within families.

Order	Species	Effective dose (Gy)	Reference
Lepidoptera	*Ephestia cautella,* Tropical warehouse moth	300	Cogburn *et al.* (1973)
	Sitotroga cerealella, Angoumois grain moth	600	Ignatowicz (2004)
	Plodia interpunctella, Indianmeal moth	350	Zolfagharieh (2004)
Coleoptera	*Sitophilus granarius,* Granary weevil	100	Ignatowicz (2004)
	Tribolium castaneum, Rust red flour beetle	120	Hayashi *et al.* (2004)
	Lasioderma serricorne, Cigarette beetle	125	Ignatowicz (2004)
	Prostephanus truncatus, Larger grain borer	300	Ignatowicz (2004)
	Oryzaephilus surinamensis, Sawtooth grain beetle	120	Tuncbilek (1997)
	Acanthoscelides obtectus, Bean weevil	60	Ignatowicz (2004)
Acarina	*Acarus siro,* Flour mite	250	Burkholder *et al.* (1966)

Fruit flies

Tephritid fruit flies comprise the group of most concern to phytosanitation, thus have been the subject of more quarantine treatment and SIT research than any other group. Irradiation phytosanitary research with this group has been comprehensively reviewed by Hallman (1999) and Hallman and Loaharanu (2002), and we will not cover Tephritidae in detail in this chapter except for new developments since 2002.

Because Tephritidae is such an important group and receives so much research effort the differing researchers and methodologies used are prone to yield divergent results. The Mediterranean fruit fly, being the most researched species of this highly researched group, is illustrative of this confusion (Table 9.3). It is generally accepted that the measure of efficacy for phytosanitary irradiation of eggs and larvae of Tephritidae (the stages present in fruit) is prevention of the emergence of adults capable of flight. The literature estimating the dose required to achieve this objective for Mediterranean fruit fly in fruit shows two peaks, one at 70–100 Gy and the other at 200–225 Gy. There are two studies showing minimal effective doses considerably higher than 225 Gy, but they can probably be dismissed as suffering from post-treatment re-infestation or other experimental problems (Hallman, 1999).

Two studies used Mediterranean fruit fly as third instars reared in diet and then inserted in fruit 24–30 h before treatment (Table 9.3). Any phytosanitary treatment infestation that differs significantly from the natural situation should be tested for relative tolerance to the natural situation. If the semi-artificial technique results in an increase in pest tolerance it would not be of phytosanitary concern, although the treatment may be harsher on the commodity and may cost more than need be. But if the semi-artificial infestation increases susceptibility

Table 9.3. Minimum ionizing radiation dose to prevent adult emergence from Mediterranean fruit fly third instars in fruit; studies listed in chronological order.

Dose (Gy)	Fruit	Reference
225	Papaya	Seo *et al.* (1973)
> 200	Orange	Fésüs *et al.* (1981)
~ 80	Mango	Potenza *et al.* (1989)
~ 80	Mango	Raga (1990)
~ 80	Peach	Arthur *et al.* (1993a, b)
~ 70	Grapefruit	Raga (1996)
~ 200	Orange	Adamo *et al.* (1996)
40[a]	Peach, orange	Mansour and Franz (1996)
150	Mango	Bustos *et al.* (2004)
100[a]	Papaya	Follett and Armstrong (2004)
100	Mango	Torres-Rivera and Hallman (2007)

[a]Fruit infestation involved rearing larvae in diet and inserting them into fruit 24–30 h before treatment.

phytosanitary security may be jeopardized. In the case of irradiation, hypoxia reduces radiosusceptibility of organisms (Hallman and Hellmich, 2007), and tephritid immatures inside the hypoxic atmosphere of fruit seem to benefit with increased tolerance (Hallman and Worley, 1999). Lack of hypoxic protection may explain why 40 Gy prevented Mediterranean fruit fly adult emergence in > 100,000 third instars reared in diet and placed in peaches and oranges 30 h before irradiation (Mansour and Franz, 1996). However, the same technique was used by Follett and Armstrong (2004) in papayas, and a dose of 100 Gy was required to prevent adult emergence. This dose seems to be the most likely minimum dose for quarantine security against this important pest, although it might even be lowered a little yet (Torres-Rivera and Hallman, 2007). Perhaps a hypoxic atmosphere was easier to achieve and maintain in papayas after artificial infestation compared with peaches and oranges, making work with papayas akin to natural conditions at least in the way done by Follett and Armstrong (2004).

Tephritids that undergo diapause do so as phanerocephalic pupae, and some may be in this stage for up to 3 years. Therefore, evaluating irradiation efficacy on adult emergence as is done for non-diapausing tephritids is not feasible for diapausing ones. Hallman (2004b) evaluated irradiation efficacy against diapausing apple maggot, *Rhagoletis pomonella*, by opening the puparia and noting development of the phanerocephalic pupal stage. Almost 38,000 third instar apple maggots were irradiated with none developing to the phanerocephalic stage. This technique allows for quick feedback during the same year for univoltine species that are difficult to rear in captivity, such as blueberry maggot, *Rhagoletis mendax*. A dose based on prevention of the phanerocephalic pupal stage should be of higher security than one based on prevention of adult emergence because the former stops development earlier and insects generally increase in tolerance as they develop.

Since at least 1991, the usefulness and possibility of a single generic, default radiation dose that would serve as a phytosanitary treatment for all tephritids on all hosts has been recognized and offered as 150 Gy (ICGFI, 1991). Although the proposal looked intuitive based on a number of research studies coordinated through the International Atomic Energy Agency (ICGFI, 1992) a number of other studies indicated that 150 Gy might not be sufficient for some tephritids (Hallman, 1999). However, Hallman and Loaharanu (2002) examined the literature on phytosanitary irradiation of tephritids and concluded that studies indicating that doses > 150 Gy might be needed could be dismissed for various reasons and argued that a generic, default dose of 150 Gy for all tephritids on all hosts was justified. Later other studies found that doses ≤ 150 Gy sufficed for some of the species where previous studies had indicated higher doses (Follett and Armstrong, 2004; Torres-Rivera and Hallman, 2007). The USA has accepted the generic dose of 150 Gy (APHIS, 2007).

Lepidoptera

Probably the next major quarantine pest group after tephritid fruit flies are the lepidopterous borers of the Noctuidae, Pyralidae, Tortricidae and associated families. Quarantine pests in the genera *Acleris*, *Acrobasis*, *Cryptophlebia*, *Cydia*, *Diatraea*, *Grapholita*, *Helicoverpa* and *Spodoptera* are in these families. The order Lepidoptera requires the highest radiation doses for phytosanitary treatment of any insects. The generic dose of 400 Gy approved in the USA for insects does not include pupae and adults of Lepidoptera (APHIS, 2007).

One of the chief species of these is the codling moth, *Cydia pomonella* (Tortricidae). It is found in much of the world attacking pome fruits and walnuts, but is not found, or is at least under regulatory control, in key importers of its hosts, such as Japan and Taiwan. Several studies have been done concerning phytosanitary irradiation of codling moth, but the one that gathered enough information to establish a dose was Mansour (2003) when he treated 100,000 fifth instars with 200 Gy and no adults emerged.

A dose of 250 Gy was found to provide quarantine security against two species of *Cryptophlebia* on fruit in Hawaii (Follett and Lower, 2000). It is possible that a lower dose would suffice; 250 Gy was used because at the time it was the minimum absorbed dose for phytosanitary control of tephritid fruit flies on some of the same fruit. Currently the dose for fruit flies is 150 Gy, but hosts of *Cryptophlebia* must still use 250 Gy. It is good practice to seek the lowest possible efficacious dose in research on any phytosanitary treatment regardless of immediate application.

Hallman (2004a) exposed 58,779 fifth instar oriental fruit moth, *Grapholita molesta*, to 195–232 Gy and prevented adult emergence although 1% of the larvae pupated. At 171–197 Gy 0.006% of fifth instar oriental fruit moth developed to adults of normal appearance.

When 30,282 sweetpotato vine borer, *Omphisa anastomosalis* (Pyralidae), pupae were irradiated with 135–148 Gy, no F_1 adults were produced (Follett, 2006a). This research, plus that with two weevils, allowed for irradiated sweet potatoes to be shipped from Hawaii to the US mainland.

Weevils

Weevils (Coleoptera: Curculionidae) are a group of important quarantine pests similar to tephritid fruit flies in feeding habit in that they feed occultly in fresh commodities. However, weevils pupate and emerge as adults inside hosts, so all stages may be found in shipped commodities. The first example of commercial use of an irradiation phytosanitary treatment against an adult quarantine pest was against sweetpotato weevil, *C. formicarius elegantulus*, on sweet potatoes shipped from Florida to California (Hallman, 2001a). The target dose in the research was 150 Gy, but because the highest dosimeter readings were 165 Gy the latter was chosen as the minimum absorbed dose for commercial application. Follett (2006a) later demonstrated that 140 Gy would suffice for sweetpotato weevil when 62,623 adults were irradiated between 125–140 Gy with no F_1 adults produced.

Another weevil that is a quarantine pest on sweet potatoes is the West Indian sweetpotato weevil, *Euscepes postfasciatus*. A minimum absorbed dose of 145 Gy prevented the F_1 adults when 60,000 parent generation adults were irradiated with 130–145 Gy (Follett, 2006a).

The plum curculio, *Conotrachelus nenuphar*, occurs in two somewhat reproductively incompatible strains: the northern, which is univoltine and undergoes obligate diapause, and the southern, which is multivoltine and undergoes facultative diapause. The southern strain was found to be more radiotolerant than the northern one. A dose of 92 Gy applied to 25,000 adults of the southern strain was found to prevent production of F_1 late larvae (Hallman, 2003).

A phytosanitary dose of 300 Gy against mango seed weevil was established in the USA (APHIS, 2007), but this does not mean that this is near the minimum absorbed dose necessary to control this insect. Follett (2001) proposed a dose of 100 Gy and doses in the 80–165 Gy range have been found to control weevils (Hallman, 2001b). However, the dose was set at 300 Gy because APHIS (2002b) cited research that indicated a dose in that range 'was necessary to effectively control the [mango seed] weevil'. A careful study with a large number of mango seed weevils is recommended to determine a more accurate dose.

There is one caution in doing research with weevils that emerge as adults deep inside fruit. The atmosphere inside intact fruits may be hypoxic and weevils irradiated under hypoxic atmospheres may have increased tolerance to the effects of irradiation (Hallman, 2005). Phytosanitary treatments should not be done in vitro with cryptically infesting pests unless testing shows there to be no difference between that and in vivo testing.

Hemiptera

A small but growing body of irradiation literature is accumulating around this large insect order that contains many important quarantine pests. It includes aphids, whiteflies, scale insects, mealybugs, leafhoppers, planthoppers, shield bugs, seed bugs and flower bugs. Usually, all stages of these hemimetabolous insects may be present on their shipped host commodities, making prevention of successful reproduction of the adult stage the only reasonable end point to phytosanitary irradiation. Where the line is drawn for prevention of reproduction is open to discussion. Some authors propose that preventing oviposition by F_1 generation adults should be acceptable. However, regulatory agencies may be reluctant to accept a treatment that allows for one generation to complete itself before total efficacy is achieved.

In a small-scale test, 200 Gy did not prevent Comstock mealybug, *Pseudococcus comstocki*, from reproducing, although F_1 females did not lay eggs (Dohino and Masaki, 1995). Jacobsen and Hara (2003) found that the dose to ensure quarantine security of pink hibiscus mealybug, *Maconellicoccus hirsutus*, was > 100 Gy and ≤ 250 Gy. Adults of greenhouse whitefly, *Trialeurodes vaporariorum*, irradiated with 50 Gy laid fewer eggs than controls and none hatched (Calvitti *et al.*, 1997). Adult green scale, *Coccus viridis*, was prevented from reproducing beyond the crawler stage with 250 Gy, the lowest dose tested (Hara *et al.*, 2002). A dose of 144–148 Gy applied to 8151 adult coconut scale resulted in 351 F_1 adults that produced no eggs (Follett, 2006b). Likewise, a dose of 128–149 Gy applied to 35,424 adult female white peach scale, *Pseudaulacaspis pentagona*, with or without eggs, resulted in 2165 F_1 second-stage nymphs and no F_1 adults with eggs (Follett, 2006c).

Thrips

Thrips are a group of important quarantine pests widely found on vegetables, cut flowers and foliage, and some fruits. Few significant studies with irradiation have been done. The dose to adults of two species to prevent reproduction was > 200 Gy and ≤ 400 Gy (Dohino *et al.*, 1996). A dose of 250 Gy, the lowest tested, prevented reproduction of yellow flower thrips, *Frankliniella schultzei* (Yalemar *et al.*, 2001).

Like hemipterans, some thrips are vectors of viral diseases of plants. To achieve quarantine security where this is an issue may require acute mortality which requires higher doses that could result in injury to fresh commodities. No studies of which we are aware consider the effect of irradiation on the ability of a vector to transmit a disease-causing organism. If irradiation reduces probing and feeding at modest doses that do not necessarily result in acute mortality, irradiation might still be a useful treatment against vectors.

Other pest groups

Irradiation phytosanitary dose ranges for the above and other groups are presented in Table 9.4. Hallman (2000) presents more detailed information on some of them. In general, lepidopterous and mite pests require the highest doses. Using irradiation as a phytosanitary treatment against nematodes on fresh commodities, except perhaps on potatoes (Karnkowski and Ignatowicz, 1999), does not seem practicable owing to the high dose required to prevent them from reproducing. However, irradiation could be used on logs to sterilize nematodes.

Table 9.4. Minimum absorbed dose ranges that might achieve quarantine security of several pest groups in order of increasing radiotolerance (Source: Hallman, 1998).[a]

Pest group	Measure of efficacy	Dose (Gy)[b]
Aphids and whiteflies	Prevent reproduction of actively reproducing adult	50–100
Seed weevils (Bruchidae)	Prevent reproduction of actively reproducing adult	70–100
Fruit flies (Tephritidae)	Prevent adult emergence from last instar	50–150
Weevils (Curculionidae)	Prevent reproduction of actively reproducing adult	80–150
Thrips	Prevent reproduction of actively reproducing adult	150–250
Borer larva (Lepidoptera)	Prevent adult emergence from last instar	150–250
Scale insects and mealy bugs	Prevent reproduction of actively reproducing adult	150–250
Borer pupa (Lepidoptera)	Prevent reproduction from late pupa	150–350
Mites	Prevent reproduction of actively reproducing adult	200–350
Nematodes	Prevent reproduction of actively reproducing adult	4000

[a] This table first appeared in Hallman (1998) and has been updated variously since.
[b] The upper dose in each range would be a reasonable guess for a generic default dose for each group. The precise measure of efficacy should be detailed for specific cases.

Arthropod pests of stored products

Disinfestation of stored products with low dose irradiation has not been used widely to date but with the phase-out of methyl bromide it can be expected to increase. High doses of irradiation have been used for many years for microbiological disinfection of high-value stored-product commodities such as spices (ICGFI, 1991) and these doses would also achieve disinfestation of arthropods, causing acute mortality. Electron beam irradiation has been used for the disinfestation of grain (Zakladnoi *et al.*, 1981) and is probably the most appropriate radiation technology for this purpose. However, currently, there appear to be logistical problems including the throughput required to achieve disinfestation at the loading rates currently employed in commercial shipping. It would be a possible replacement for methyl bromide when used for fast disinfestation of grain diverted to adjacent special treatment facilities when found to be infested during loading of ships. Sea transit times for grain are often long enough for mortality of pests resulting from gut cell injury to be effective in achieving pest free grain at arrival, as well as preventing multiplication in transit. This might require amendment of legislation of countries such as Australia, which require grain to be free of live pests at export (Pheloung and Macbeth, 2002). Efficacy levels required against regulated non-quarantine pests of stored products for phytosanitary rather than quarantine purposes can be lower than 99.9968% for some markets.

Researchers studying the effects of irradiation have made extensive use of some species of stored-product insects, especially *Tribolium* spp. This has resulted in a considerable volume of scientific literature in IDIDAS (FAO/IAEA, 2007) on genetic and physiological effects of irradiation on insects. *Tribolium* spp. have been used as one of the study insects in investigation of low energy electron treatments by Japanese and Bangladeshi researchers (Hayashi *et al.*, 2004). Coleopterous pests of stored products are susceptible to relatively low disinfestation dose levels, with beetles of the family Bruchidae (e.g. *Acanthoscelides* sp.) possibly the most susceptible at less than 100 Gy (Table 9.4). However, the stored-product moths especially the Angoumois grain moth, *Sitotroga cerealella*, can require up to 600 Gy to ensure infertility (Ignatowicz, 2004). Doses reported in IDIDAS for inhibition of development and sterility of adult stored-product moths exhibit considerable variation (FAO/IAEA, 2007) and further research seems justified.

Commodity Tolerance

Some of the earlier research on commodity tolerance used doses designed to provide acute mortality of the pest, and thus, were higher than needed for quarantine security. Results with some commodities have been highly variable indicating that uncontrolled factors, such as precise level of maturity, may not have been controlled. Commercial interests considering irradiation as a phytosanitary treatment should use previous research only as an indication of possible tolerance and do enough applied research themselves to be satisfied with the outcome of product quality. Often, commodities that are harvested in a state not immediately ready for use should be harvested later to achieve optimum quality following irradiation as irradiation may interfere with the ripening process.

Much research on the effect of irradiation on cut flowers has been done in Japan. Radiation damage to cut flowers ranges from early petal fall, with consequent reduced shelf life, to disfiguring phytotoxic injury. Damage varies from species to species (Kikuchi *et al.*, 1999). The more mature the bloom, the more tolerant it is likely to be, although radiation may delay opening of blooms, thus increasing flower shelf life. Measures are available to increase levels of tolerance including nutrient solutions (Hayashi *et al.*, 1999). In some cases irradiation may preserve quality (Fig. 9.4).

For fresh fruit, provenance, maturity, ripeness and pre-treatment storage have been shown to influence susceptibility to injury from radiation treatment in mangoes (Boag *et al.*, 1990) and similar influences can be expected in other fresh horticultural commodities. A guide to tolerance of general commodity groups to irradiation was given by Moy *et al.* (1999). Most commercially traded fruits, vegetables and cut flowers at commercial maturity can tolerate the minimum doses required as phytosanitary treatments against arthropod pests, but not those required against nematodes and plant pathogens. Strawberries are a notable exception, with a tolerance threshold of at least 2 kGy. An important operational factor is the DUR (dose uniformity or max:min ratio) of the treatment when applied in a commercial irradiator. This ratio can be as high as a factor of three and consequent injury problems can arise where authorities add a confidence margin to the experimentally demonstrated minimum, a common practice in the past for fumigants and other pesticides. A prudent practice is to check for susceptibility to injury by testing a representative sample of the intended commodity at the intended treatment schedule in the intended facility before commencing large-scale operations.

Fig. 9.4. Shelf life of red ginger flowers prolonged by irradiation 18 days earlier at 335–422 Gy (left) compared with non-irradiated control (right).

For durable commodities including harvested grains the tolerance can be expected to be adequate for most arthropod pests. It is possible that baking quality of flour may be affected in some instances at the much higher doses required for microbial decontamination. For seeds and other commodities such as malting barley where the viability of the grain or seeds is important, some adverse effects can be expected although they may still lie within levels that can be tolerated commercially.

Research

Dosimetry results are crucial to the reporting of any phytosanitary treatment research. Often these results have not been reported with irradiation. It may be that the actual dose absorbed by the test organism sometimes exceeded the dose reported. Where doses are reported only as single values these are usually what is estimated to be achieved at the mid-plane of the product from a parallel source and the actual dose at any other point will be higher or lower. Even if the product is irradiated from each side the dose estimate will not be as precise as if it were estimated from a number of dosimeters distributed throughout the product. Consequently, experimental doses should be measured and expressed as a range with the maximum experimental value becoming the minimum operational dose.

Research on irradiation as a phytosanitary treatment has unique complications compared with research on any other treatment presently used commercially because the measure of efficacy is rarely a dead pest, as it is with all other commercial treatments. Because there is no independent confirmation of efficacy, such as all quarantine pests dead upon inspection, research on irradiation phytosanitary treatments must be done correctly from the start. There will be no second chance for post-treatment inspection to indicate when a treatment fails as has happened with every other major treatment. Also, because the measure of efficacy is not dead pests, but prevention of development past a specific stage or prevention of reproduction, the experimental organisms must be kept under good conditions until they finally die, and the non-irradiated controls must live and reproduce within normal expectations. This requires greater biological skills than research with treatments that result in mortality soon after the treatment is finished.

Operational Standards

In 2002 the USDA-APHIS enacted a final rule for regulation of 'Irradiation phytosanitary treatment of imported fruit and vegetables' (APHIS, 2002b). This rule, which is essentially a de facto standard, was enacted following consideration of extensive public comment from within the USA and overseas. Subsequently, minor additional changes were made in conformity with the principle of updating as new information becomes available. These regulations were based on a concept of generic dosages with respect to commodities and listed scheduled dosages against listed pests as discussed earlier in this chapter. The regulations (inter alia) currently deal with:

- Approved doses.
- Location of facilities.
- Compliance agreement (inside/outside the USA).
- Certification of facilities.
- Monitoring of interagency agreements.
- Packaging.
- Dosimetry systems.
- Records.
- Request for certification and inspection of a facility.
- Denial and withdrawal of certification.

In 2003 the IPPC (2007) published *Guidelines for the Use of Irradiation as a Phytosanitary Measure*, which is essentially a code of practice or advisory standard. This standard, which borrowed from previous Food and Agriculture Organization (FAO)/IAEA documents, can be used as a template for the development of a national standard by any country or it can be specified as it stands. These guidelines detail:

- Definition of the regulatory authority.
- Treatment objective including required efficacy.
- Application of the treatment.
- Dosimetry including calibration of components of the dosimetry system, dose mapping and routine dosimetry.

- Facility approval.
- Integrity of the phytosanitary system including phytosanitary security measures, labelling and verification.
- Documentation including facility records and traceability.
- Inspection and phytosanitary certification by the regulatory authority including export inspection, phytosanitary certification, import inspection and verification methods for treatment efficacy in export and import inspection.
- A protocol for research to develop treatments.

Other countries have approached the approval of quarantine treatments by irradiation in different ways. Australia and New Zealand have a bilaterally agreed procedure under which their general food irradiation standard A17 covers phytosanitary applications on a case-by-case basis (FSANZ, 2000).

Conclusions

The history of research on ionizing irradiation as a phytosanitary treatment is almost as long as the history of any other phytosanitary treatment. However, it has not been used on a continuing basis until 1995. As of this writing it is used for modest interstate shipments among the US states, mangoes and papaya from Australia to New Zealand and for a few shipments of mangoes from India to the USA. Other countries seem poised to begin using it. Irradiation has the potential to solve many phytosanitary problems and is the most tolerated treatment by fresh commodities in general. Drawbacks are the initial price to set up a facility plus the regulations surrounding management of the facility. It will not be as widely available as are cold, heat and fumigation facilities, resulting in logistical challenges and possible added costs of transportation. Locating irradiation facilities at ports that commodities already use would lessen this logistical burden. The organic industry does not accept irradiation. This is not a science-based stance, as irradiation has not been shown to make food unhealthy in any way.

10 Disinfestation by Fumigation

A fumigant is a chemical that exists in the gaseous state at normal temperatures and pressures in sufficient concentration when applied, to be lethal to pests (Bond, 1984). Being gases, fumigants diffuse as separate molecules enabling them to penetrate deeply into products and then diffuse away after fumigation. Aerosols (see Chapter 12) as smokes, fogs or mists are sometimes mistakenly called fumigants, but they are made up of liquid or solid particles that do not diffuse and remain adhered to the surface of the product after treatment. Some liquid pesticides do have an active vapour phase. For example, dichlorvos is effective as a fog or fumigant, but has now been removed from use in some countries because of potential carcinogenic properties.

The use of toxic gases to disinfest food commodities and buildings dates from antiquity when the use of heated sulphur fumes was recorded. Modern concepts of fumigation date from the formulation and subsequent use of carbon disulphide and hydrocyanic acid in 1854 (Winks, 1984). Since then other fumigants have been developed and the technology has been a mainstay of phytosanitation for many decades.

Fumigation for insect control was comprehensively reviewed in a United Nations Food and Agriculture Organization (FAO) manual (Bond, 1984). More recent reviews cover characteristics of fumigants and past research with them (Stark, 1994), fumigation as a phytosanitary treatment for fresh fruits and vegetables (Yokoyama, 1994), responses of fumigated plants and living plant parts to fumigation for pest control (Forney and Houck, 1994), and as a technology on the wane (Banks, 1994). The US Department of Agriculture (USDA) Treatment Manual (APHIS, 2007) contains comprehensive information on fumigation for the purpose of quarantine policies and regulations as they apply to the USA for state and international users. Much of this information is of general technical interest to authorities in other parts of the world.

The practices of fumigation for phytosanitary purposes are well established but changes can be expected from time to time, especially in relation to the fumigant usage approved by both exporting and importing countries. As an agricultural chemical, legal regulation of fumigants is highly complex and liable to change so the most recent approvals should be sought from relevant national regulatory authorities before proceeding with any treatment. The maximum residue limit (MRL) is pivotal to the regulation of all agricultural chemicals. Most countries do not accept commodities with detectable residues of agricultural chemicals for which they have no nationally approved MRL so this effectively controls trade in all treated commodities.

Fumigants penetrate bulks and even commodities such as individual fruits or wood to provide effective internal disinfestation. Some are able to achieve this within a few hours. Death of the pests is not always instantaneous and may take some time following removal of the commodity from the fumigant atmosphere. A few individuals may be found moving, up to several days following fumigation. This can result in some consternation when regulatory inspectors find live but moribund pests in importations of properly fumigated commodities; inspectors prefer to find only dead pests. Indeed, conditions of sale contracts may require a phytosanitary certificate stating that no live insects were found at pre-export inspection and solutions should be negotiated when it is possible that live, but moribund, pests may be found upon inspection. This has been successfully accomplished with irradiation where live pests after treatments are the norm (Chapter 9).

During the past century, fumigants used for phytosanitary purposes have included carbon bisulphide, carbonyl sulphide, cyanide, ethylene dibromide, ethylene oxide, methyl bromide, phosphine, sulphuryl fluoride and others. Health and environmental controls have reduced current options in most countries to methyl bromide, phosphine and sulfuryl fluoride, with methyl bromide under strict reduction as a stratospheric ozone-depleting substance. Alternative fumigants have been studied, most with scant short-term promise of being approved for general use; consequently, there has been a tendency to move away from fumigants for phytosanitary purposes when other alternatives are possible (Fons, 1990).

Fumigation Facilities

Fumigation chambers can be purpose-built permanent structures (Figs 10.1–10.3) or can be temporary and constructed of gas-impermeable flexible fabric used to envelop a commodity bulk. The gas tightness of the fumigation space is a major factor in determining the initial dosage and the possible need for subsequent supplementation to ensure an effective fumigant concentration throughout the exposure period. For a slow-acting fumigant such as phosphine this can mean a dosage requirement difference factor of up to four times the dose for a well-sealed chamber.

The size and shape of the fumigation chamber determines the surface to volume ratio and this influences the potential gas loss rate. Small chambers require greater attention to gas tightness than large chambers. Gas tightness of an enclosed fumigation space should be checked before introduction of the gas by

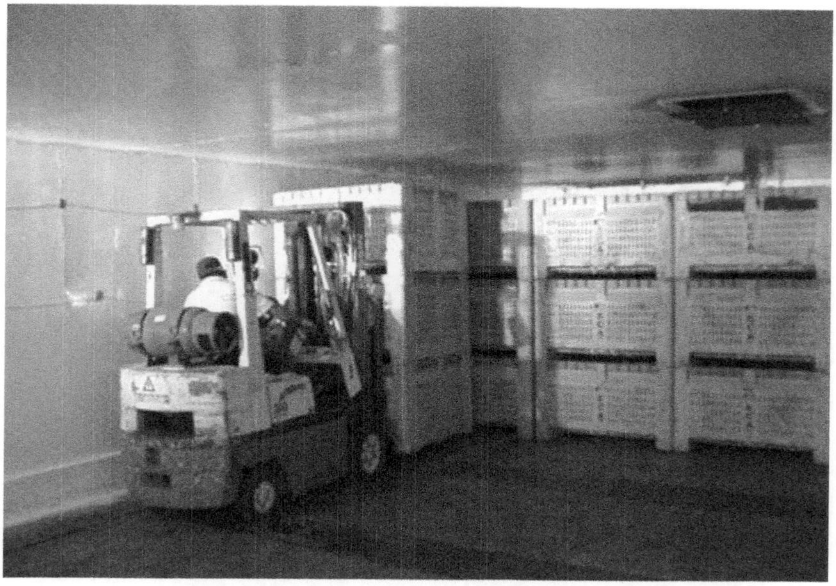

Fig. 10.1. Fruit being prepared for fumigation with methyl bromide in a sealable purpose-built chamber.

using a simple water gauge and noting the half-life of a low-pressure inflation with air or, in the case of low vacuum fumigation, gain of pressure.

Temperature of the commodity at the time of fumigation is probably the most important operational factor, affecting as it does the physical characteristics of the fumigant, especially the gas/liquid phases, and the physical activity of the gas atmosphere. The lethal effect of a fumigant on most pests depends to a large extent on their metabolic rate, which is temperature related. Adult and juvenile insects absorb most fumigants through spiracles leading to their internal respiratory system, but there can be diffusion through the cuticle at a lesser rate. Insect eggs absorb fumigants mostly by diffusion through the chorion. Due to the egg's small size the chorion is unlikely to represent a significant barrier. When eggs are cited as being more tolerant of fumigation than larvae it is sometimes because different end points are being measured. For example, Armstrong and Whitehand (2005) concluded that the egg stage of two tephritid fruit flies was more tolerant of methyl bromide than larvae, but they used hatching as the end point for egg survival, and pupariation and adult emergence, respectively, as end points for young and old larvae. In this case the eggs had only to hatch to be counted as survivors while both larval groups had to pass through several developmental stages to be counted as survivors. End points should reflect similar challenges.

Depending on ambient and commodity bulk temperatures, auxiliary equipment may be required to heat the fumigant to achieve the gaseous phase, to circulate the gas within the space, to monitor the gas concentration during fumigation and to purge the fumigant by aeration after the required exposure period

Fig. 10.2. Gassing equipment of a permanent fumigation chamber. (a) Note that the operator is using a positive-flow personal air supply as he controls the amount of gas admitted from the cylinder by weight using a heated vaporizer to ensure that the liquid methyl bromide is admitted as the gas phase. (b) A high volume extraction fan is installed to ensure rapid clearance of the gas after treatment is completed.

Fig. 10.3. Gas delivery equipment needed for a commercial fumigation with methyl bromide (black tank). The large tank is for heating the methyl bromide to ensure its gaseous form before entering the fumigation chamber (note flame at bottom). The small tank is the propane fuel source.

(Figs 10.2–10.4). Gas re-circulation can markedly reduce the total fumigation time required. Capture of the used fumigant (Leesch, 2002), where feasible, may become a future requirement for environmental reasons (EPA, 2005a).

Fumigants are subject to sorption by the surface material of the chamber as well as the product (Bond, 1984; APHIS, 2007). Sorption rate is typically high at first then gradually reduces to a slow rate. The phenomenon is temperature related and can reduce the activity of a fumigant to a significant extent, requiring concentration monitoring and supplementation in the case of some commodities. It can be temporary, resulting from physical adsorption and absorption, or permanent as a result of chemical reaction. It can relate to the surface fabric of the chamber (e.g. hardboard) as well as to the commodity (e.g. flour). APHIS (2007) lists 18 diverse examples of materials with high sorptive rates. Desorption of physical sorption is temperature related and can be accelerated by aeration. Chemically sorbed fumigant will be detectable as an inorganic residue. There is considerable latitude in the literature involving the use of the terms adsorption and absorption. Some authorities restrict the terms adsorption to physical sorption and absorption to permanent chemical sorption.

Safety and Health Considerations

Fumigants are broadly toxic to animals, and death and injury of application personnel regrettably occur. For example, each year in the recent past the US state of California recorded around two deaths and over 24 injuries due to methyl

Fig. 10.4. Setting up for re-circulation fumigation of a vertical grain storage with methyl bromide, using an axial flow fan and heating to vaporize the gas before introduction to the base of the silo.

bromide poisoning, mostly in field use (Anon., 1994). These are avoidable through protective accessories (Fig. 10.2) and strict adherence to safety procedures given on container labels. The most hazardous fumigants will often carry regulatory requirements such that they can only be used by trained and licensed operators and this requirement can be expected to extend to those which are currently unrestricted in this way.

Threshold limit values (TLV) of the fumigant in use need to be known and monitored by operators, together with safety exclusion zones and gas clearance protocols. Fumigants such as methyl bromide, not readily detected by smell, may be formulated commercially with trace amounts of chloropicrin, itself a fumigant, to enhance their passive detection at low concentrations. However, chloropicrin is not added to methyl bromide used for postharvest fumigation because of the objectionable odour and damage it may cause to fresh commodities. Technology is available for monitoring gas leaks at low concentrations from fumigation chambers, including infrared analysers, a flame colour change lamp for halides and Drager® tube air samplers for phosphine and a range of other gases.

Most fumigants will leave recognizable residues which will be subject to tolerance limits. Monitoring is an essential part of safety measures and modern analytical equipment permits analyses with levels of detection now frequently decreased from parts per million (ppm) to parts per billion. The ability to detect minute amounts of residues can cause a heightened sense of concern in users although the basic hazard levels remain unchanged. When used according to label recommendations, residues should not pose health or safety hazards, so label advice takes precedence in all handling practices. Fumigant use cannot violate label prescriptions under penalty of law.

Commodity Considerations

Fumigants, being broadly toxic, may injure fresh perishable commodities (Forney and Houck, 1994). Injury to the commodity is often caused by enzyme inhibition or membrane damage among other causes. Symptoms may be expressed in necrotic spotting, changes in odour and taste, altered senescence and increased decay. Expression and intensity of damage vary greatly for individual commodities owing to factors inherent in the application itself (e.g. dose, duration, temperature, load, humidity, atmospheric pressure) and characteristics of the commodity being fumigated (e.g. cultivar, maturity, ripeness, seasonal growth conditions, time since harvest, postharvest handling and application of coatings or other chemicals). Fumigants can reduce the shelf life of commodities without the expression of other noticeable injury. The risk needs to be assessed before a decision is made to proceed with a treatment. Sorption rates differ between commodities and can even differ among individual cultivars.

Although most pests of grain are cosmopolitan, having been transmitted across the globe with stored foods as long as humans have been migrating, they can be categorized as non-quarantine regulated pests at arrival and warrant phytosanitary action before export or, when detected, at arrival. A few stored-product pests are of quarantine concern, such as khapra beetle, *Trogoderma granarium*, and the larger grain borer, *Prostephanus truncatus*. Durable commodities such as food grains and other stored products are less likely to experience damage although multiple fumigations may lead to noticeable changes in colour, sheen or other attributes. Seeds are subject to injury by some fumigants, leading to deleterious effects on germination, emergence and seedling vigour. Moisture content can have a major influence on the susceptibility of seeds to injury, with, in general, the lower the moisture the less the likelihood of injury. The quality of grains used in malting may also be affected by some fumigants.

Measures are necessary to prevent post-fumigation re-infestation. Quarantine security can be assured by the use of insect-proof but gas-permeable packaging before treatment, a procedure for which fumigation is ideally suited despite the possibility of slower desorption. If the commodity is packaged after treatment or otherwise handled in bulk, adequate security management procedures should be put in place. Although fumigants may leave residue traces, these are too low to provide protection against re-infestation.

Some fumigants can affect fabric and fittings of chambers. For example, phosphine attacks copper fittings including electrical circuitry and liquid methyl bromide reacts with aluminium. Care must therefore be exercised to ensure that no vulnerable materials are exposed.

Fumigation Dosimetry

Dosimetry of any phytosanitary treatment is fundamental to the entire treatment process. If the dosimetry system does not give accurate and consistent measurements, little confidence can be placed in the efficacy of the treatment. Measurement of fumigant concentration commences with the initial dosage or immediately after equilibration has occurred within the chamber atmosphere. The term 'dosage' is customarily applied to the amount of fumigant introduced to the treatment chamber, while the term 'dose' is applied to the concentration × time (CT) product required to ensure mortality of the pest (Winks, 1984) and is the amount of fumigant exposure the target pest receives.

Within certain limits and depending on the decay pattern of the concentration (c), fumigant efficacy (k) can have a linear relationship with time (t) such that $k = ct$, known widely as the 'CT product'. During the fumigation period the gas concentration can be monitored using instruments appropriate to the fumigant. Multiple sampling points should be used, as it is important to know the degree of homogeneity of the fumigation atmosphere. CT product should not be used for phosphine because of the abnormally long exposure times needed (Bond, 1984).

For determination of the initial fumigant concentration and subsequent monitoring, equipment may be specified according to the precision required. Options include gas chromatography, interference refractometry, colour indicators, lamps, thermal conductivity meters, halide meters and colour change detector tubes in association with an air pump. These items of equipment are also used for operator and public safety purposes when calibrated for much lower concentrations.

Where the initial concentration of a fumigant is at higher than recommended levels for a pest, protective narcosis or stupefaction can occur resulting in lowered metabolic activity such that the eventual lethal dose is higher than it would have been otherwise (Bond, 1984). Narcosis may facilitate the genetic phenomenon of resistance but is by no means the only factor involved.

Fumigants

After ethylene dibromide was lost as a commodity fumigant for almost all of its uses in the mid-1980s and methyl bromide restrictions began to be discussed in the early 1990s the future for fumigants as phytosanitary treatments looked bleak. But in recent years there has been a resurgence of interest in treatments with modified formulations of current fumigants and formulations of new fumigants. This resurgence has been fuelled in part by the difficulty in finding replacements for ethylene dibromide and methyl bromide and the fact that fumigation is a well-accepted method that is logistically easy, effective and relatively economical to apply.

Much of the research effort with alternative fumigants and formulations is summarized in proceedings of the Annual International Research Conference on Methyl Bromide Alternatives and Emissions Reductions (MBAO, 2007). The proceedings since 1994 are available on the website. Reports are abbreviated and mostly preliminary, but do give researchers and other interested parties updates on developments and possibilities in the field. Although the conference contains the words 'methyl bromide alternatives' some of the phytosanitary research summarized may not be alternatives in the sense that methyl bromide was never used for the problem to begin with. This conference may be the most relevant single annual meeting for phytosanitary workers covering other alternatives besides fumigants. A major part of the programme deals with pre-sowing or pre-planting replacements for methyl bromide.

Methyl bromide

Methyl bromide is currently the predominant fumigant for phytosanitary purposes. It was widely used over several decades as a soil fumigant against soil pathogens, weeds and pests in the field, nurseries and glasshouses; for the disinfestation of grains and other stored products and packaging; for the disinfestation of fruits, vegetables, cut flowers and seeds; and for the devitalization of cut flowers and weed seeds. Now it is considered a significant stratospheric ozone-deleting substance so increased regulation of the fumigant must lead to lowered use in the future, although it is generated and absorbed naturally in seawater systems.

Methyl bromide was developed in the 1930s as a phytosanitary fumigant tolerated by many fresh commodities at the doses necessary to kill quarantine pests (Bond, 1984). Its use is facilitated by the relatively simple technology required to achieve highly efficacious treatments (Figs 10.1–10.4). Broad applicability led to its adoption for many uses without the rigid experimental testing and determination of dosages required of many subsequent treatments. However, from the late 1950s to the mid-1980s ethylene dibromide, which was superior to methyl bromide for internal pests such as fruit flies, replaced many phytosanitary uses of methyl bromide. In 1984 ethylene dibromide was banned in the USA as a presumptive carcinogen and mutagen (EPA, 1984), and methyl bromide regained its pre-eminence as the fumigant most used for phytosanitary purposes. Other countries soon followed suit, although as late as the late 1990s ethylene dibromide was used for internal quarantines in a fruit fly eradication effort in Australia (Hallman, 2002).

Although methyl bromide is not as broadly efficacious as a few other fumigants, such as ethylene dibromide and hydrogen cyanide, it has many logistical and operational advantages due in part to the availability of the chemical as a liquid with a long shelf life in containers as small as 450 g. The relatively low boiling point of methyl bromide enables it to be transported as a liquid but used as a gas at low temperatures. Its relatively high molecular weight makes it easier to contain in a fumigation space than lighter fumigants, although it still is easily and rapidly sorbed at normal atmospheric pressures. Methyl bromide fumigation may be done at temperatures as low as 4.5°C which is lower than most other fumigants used against various arthropod pests on deciduous fruits (Bond, 1984).

Methyl bromide is non-flammable in air although in the liquid form it reacts with aluminium in the absence of oxygen to produce methyl aluminium bromide that will ignite spontaneously when exposed to oxygen. Therefore, methyl bromide fumigation chambers should not contain any aluminium tubing in the fumigant intake system nor should the fumigant be stored in containers that have aluminium parts.

Although treatment schedules for methyl bromide tend to be somewhat specified by individual pests or pest categories, hosts and commodities, generic grouping is done as well (Bond, 1984; APHIS, 2007). Treatment schedules have been extrapolated to some specific pests and commodities for which efficacy or tolerance data are not available. Typical dosages for pests of fresh fruits range from 16 g/m^3 at 32°C to 54 g/m^3 at 4.5°C. For fruits, vegetables and cut flowers usual exposure times are 2–4 h. Internal feeding pests such as tephritid fruit flies or Lepidoptera larvae require higher doses compared to external feeding pests such as thrips, mealybugs, mites, aphids and scale insects. Dosage levels up to double those for fresh commodities are used for some pests of grains and stored products. Exposure time required can be up to 24 h although times of 12 h or less can be achieved for grains by the use of fumigant re-circulation techniques (Fig. 10.4) or under vacuum fumigation. For logs the dosage required can be up to 240 g/m^3 and the exposure time as long as 72 h (APHIS, 2007). As with other chemicals used on food, most countries have regulated MRL for methyl bromide considered as inorganic bromide and calculated as part of total bromides.

Methyl bromide fumigation can be combined with a cold treatment (Chapter 7) in some cases (FAO, 1984; APHIS, 2007). The severity of either treatment alone is insufficient to provide quarantine security. The reasoning for this combination treatment is that some commodities may not tolerate a more severe cold or methyl bromide treatment alone or that the cold can be done in part in transit but the transit time is insufficient for it as a stand-alone treatment.

The Montreal Protocol arising out of the Vienna Convention of the United Nations Development Programme sets phase-out targets for the use of atmospheric ozone-depleting substances including methyl bromide (UNIDO, 2007). For developed countries, use levels were established based on labels used on 1 January 1995 and a phase-out date of 1 January 2005 was established with exemptions permitted subject to annual review (EPA, 2005a). Most postharvest phytosanitary uses are exempted, although this status could change. For less developed countries, the use levels and phase-out dates are 1 January 2002 and 1 January 2015, respectively.

Most methyl bromide usage is related to soil treatment for plant pathogens and nematodes, for which wider-based options may be available, so major reductions were achieved early. However, for postharvest quarantine purposes practical alternatives are not always readily available even in developed countries. In recognition of this, applications for critical use exemptions can be made to their national organizations responsible for overseeing the phase-out, such as in the USA, the US Environmental Protection Agency (EPA) which administers their Clean Air Act (EPA, 2005a). Long-term usage might be permitted if technology to capture and re-use the fumigant can be developed and approved by environmental authorities. Some research focuses on capture of the gas using a carbon filter which is then thermally or catalytically treated to convert the methyl bromide to water, carbon dioxide and hydrogen bromide and eventually to bromine as a stock chemical (Leesch, 2002). In this way 95% recovery is claimed possible, thus minimizing atmospheric contamination. Other technology is also under development using scrubbing methodology (UNIDO, 2007).

Measures including reduced production of methyl bromide and preventing its loss to the atmosphere are likely to increase costs and have stimulated a broad-based search for alternative strategies (MBAO, 2007).

Phosphine

Phosphine fumigation has traditionally been done with solid formulations, such as pellets and tablets. Recent gaseous formulations have allowed for a broader diversity of application methods and commodities. Gaseous formulations have allowed for the removal of by-products of phosphine release from solid formulations that may prove to be a nuisance with fresh commodities.

Phosphine has been shown to be vulnerable to the development of resistance in some species of stored-product beetles especially the red flour beetle, *Tribolium castaneum*, the rice and granary weevils, *Sitophilus* spp. and the lesser grain borer, *Rhyzopertha dominica* (White and Lambkin, 1990; Zettler and Cuperis, 1990). The resistance status of pest populations should be taken into account when prescribing fumigation treatments against these pests (Ducom, 2005).

Phosphine is not generally used for plant propagative materials or fresh produce because of the damage usually caused to living commodities other than dormant seeds at the concentrations and times required for short shelf-life commodities (Hatton *et al.*, 1982).

Fumigation with solid formulations

Phosphine (hydrogen phosphide) is widely used for phytosanitary purposes on seeds, grains and other stored products. The already extensive literature on it has expanded recently as it receives attention as an alternative to methyl bromide. At usual concentrations it requires several days to achieve complete pest control. It is more volatile than other common fumigants such as methyl bromide, at operational temperatures and consequently requires greater attention to sealing, or compensation for loss. The MRL for phosphine tends to be set at the level of analytical quantitation (currently a nominal 0.01 mg/kg in Australia for a range of commodities including a number of fruits and vegetables) as a measure of caution given current uncertainty about possible deleterious effects. The level of detection is dependent on the methodology and needs to be defined.

Commercial phosphine is formulated as aluminium or magnesium phosphide. Tablets (3 g), pellets (0.6 g), plates (206 g), sachets (125 g) or mats of these preparations make for flexibility and safety of use as release is relatively slow due to reliance on atmospheric moisture to facilitate the release reaction. Contact with free water can produce a dangerously rapid rate of release coupled with possible fire or explosion. Phosphine is also available as cylinders of the gas, in various formulations some with carbon dioxide, others undiluted, but usage in these forms could be subject to more intensive regulatory control than the safer solid formulations.

Aluminium or magnesium phosphide is not normally used at relative humidities of < 40% or temperatures < 5°C (Bond, 1984). This does not present many problems in practice because stored-product insects are curtailed naturally by these physical conditions. For grain bulks, the relative humidity at a given commodity moisture content can be obtained from tables on equilibrium moisture contents (EMC) of grains. Recommended dosage rates for phytosanitary disinfestation of grain with phosphine range from 1.5 to 5 tablets/t or the equivalent in other formulations depending on the gas tightness of the fumigation structure. Dose can be specified as a CT figure, for example 6 g/m^3 for 168 h for disinfestation of baled hay (APHIS, 2007). Bond (1984) listed treatment dosages of 1–2 g/m^3 depending on the commodity with minimum exposure times of 4–7 days, depending on temperature and possibly humidity, to ensure that all phosphine had been released before further handling. Dosage can be related to tabulated data on the amount of phosphine released by various formulations (APHIS, 2007). Phosphine tablets or pellets can be introduced to grain bulks by the use of a probe, or they can be added to a moving grain stream as a sealable storage structure is filled. Tablets, pellets or plates can be placed on the surface of a grain bulk to be sheeted. In-transit fumigation phosphine has been proposed and at times used, but few countries or international transport organizations would favour or condone it, even with stringent controls, because of risks to ships' crews and stevedores (Low *et al.*, 2003).

Fumigation with gaseous formulations

Formulations of 2% gaseous phosphine with the remainder carbon dioxide offer greater potential for this fumigant and have the advantage for fresh fruits and vegetables of absence of the aluminium or magnesium powder residue. This formulation was developed in the 1980s for grain fumigation. Winks (1993) describes a continuous slow release technique into a pressurized distribution system at the base of a vertical storage with good gas retention properties. The combination of phosphine and carbon dioxide was later tried successfully against surface pests on cut flowers (Williams and Muhunthan, 1998) on which it is presently registered for use in Australia. Williams *et al.* (2000) achieved 99.9966% control with the formulation at 25°C for 48 h against larvae of Queensland fruit fly, *Bactrocera tryoni*, in oranges. The phosphine concentration was measured at 1.67 and 0.14 g/m^3 at the start and at the end of the fumigation, respectively. Their measure of control was prevention of pupariation from larvae that emerged from fumigated fruit and fell into vermiculite below. Some doubt remained about the ability of the fruit to withstand the conditions of fumigation. Preliminary tests detected no adverse effects of the fumigation on orange quality and taste.

Soma *et al.* (2002) reported control of tetranychid mites on Japanese pears without injury to the fruit in large-scale trials using 1.5 g/m^3 for 24 h of 'generated' phosphine at 15°C. Brash *et al.* (2002) reported control of *Thrips tabaci* on onions with a 2% phosphine formulation in carbon dioxide after 2 days for adults and 3 days for eggs. Against adults, 40% carbon dioxide without the phosphine was effective although it did not prevent egg hatch.

Research on fruit pests with formulations of gaseous phosphine in air were reported at the 2005 Annual International Research Conference on Methyl Bromide Alternatives and Emissions Reductions. Horn *et al.* (2005a) summarized a technique to apply gaseous phosphine in air at cool temperatures said to be used commercially in Chile. Depending on the fruit, treatments are done between −1.5 and 15°C with 1–3.5 g/m^3 of phosphine for 24–72 h. The authors claim control of a number of pest species, some of which would be relevant for phytosanitary requirements but provided no efficacy data. It is not clear how much of the efficacy of these treatments would be attributable to cold. Horn *et al.* (2005b) found that for the quarantine mite pest of a number of fruits and ornamentals, *Brevipalpus chilensis* (family Tenuipalpidae), control with phosphine required 72 h exposure at a dosage of 1500 ppm and temperature of 6°C followed by < 11-days cold treatment at 0°C. At the same conference, Ducom (2005) advised against the use of phosphine against stored-product pests at temperatures below 10°C because of long times for mortality.

Sulphuryl fluoride

Sulphuryl fluoride was developed for structural fumigations to control drywood termites in the USA. It is approved for disinfestation of non-food items against ticks (Acarina) and for a number of food items in the USA such as grain, dried fruit, meat, cheese, coconut, cottonseed, groundnut, ginger and legumes (EPA, 2005b; APHIS, 2007). The prescribed dosages range from 32 g/m^3 at 21°C to 48 g/m^3 at 4°C for 24 h but fumigations should not be done below 10°C. Sulphuryl fluoride fumigation shows promise for killing larvae of Lepidoptera in walnuts and almonds, especially at low atmospheric pressure (Zettler and Leesch, 2000). Reporting in an abstract, Mack and Barak (2006) found the fumigant to be 'highly effective' against two snail species that were intercepted in imports of ceramic tiles. With hardy commodities such as tiles the risk of damage caused by the treatment is usually very low. Sulphuryl fluoride did not control red scale, *Aonidiella aurantii*, or first instars of Mediterranean fruit fly, *Ceratitis capitata*, in lemons at dose levels tolerated by the fruit (Obenland *et al.*, 1998; Aung *et al.*, 2001).

Dichlorvos

Snelson (1987) provided detailed technical information on dichlorvos as a disinfestant or protectant for seeds or grain. Bond (1984) reviewed its action and use as a fumigant. Dichlorvos is generally formulated as an emulsifiable concentrate but can be packaged with propellant for use as an aerosol or impregnated in resin strips for slow release in an enclosed space. For seeds and grain it can provide disinfestation for phytosanitary purposes by admixture of concentrate diluted with water and applied according to manufacturers directions as a spray at 1 l/t (White and Blackett, 1988). Subsequent storage in a sealed space can improve the efficacy of a treatment. This can provide disinfestation in as little as 24 h but it is more effective against stages external to grains or seeds than to internal stages. It confers only limited residual protection.

Dichlorvos has been used for the in-transit disinfestation of cut flowers (Heather, personal observation) by the inclusion of the resin strip formulation within shipping packages. It is not generally phytotoxic, although slight discoloration of chrysanthemum flowers and burning of foliage has been noted (Bond, 1984). A check should be made to determine whether such usage may require approval before it is used for a specific purpose. There are no known approvals for its use with fruit in this way. Such approvals could be expected to be much more stringent than for cut flowers although specific Codex-based MRL ranging from 0.02 to 5 mg/kg are in place including one of 0.02 mg/kg for mushrooms (IPSAPH, 2007). Use of dichlorvos in the USA is being further restricted voluntarily by the manufacturer (EPA, 2007). It is listed as 'possibly carcinogenic to humans' (WHO, 2007).

Ozone

Ozone (O_3) is of value in the stratosphere because it filters the shorter wavelengths of ultraviolet light, which can cause ionizing damage to living cells. That, of course, is the crux of the problem with methyl bromide; significant amounts rise to the stratosphere and catalyse the reaction to turn ozone back into oxygen. However, in the lower atmosphere ozone is considered a pollutant formed by the reaction of ultraviolet light on hydrocarbons and nitrogen oxides in the air. Ozone reacts directly with organic double bonds and as it breaks down gives rise to oxygen free radicals, which damage organic molecules. It is this biological property of ozone that makes it potentially useful as a fumigant.

Leesch *et al.* (2003) used 1% ozone and carbon dioxide at several percentage points in a vacuum (about −35 to −40 kPa) for 2 h as potential treatments against different pests. The advantages are that the ozone is easily destroyed after fumigation to leave none to enter the lower atmosphere, it is generated on site to avoid transportation costs and hazards, and it is efficacious against several pests. Disadvantages include that its strong oxidizing properties damage metals, it does not penetrate as well as other fumigants, and it injures some commodities. Researchers continue to develop proposed fumigation schedules using ozone, especially with other factors such as carbon dioxide and vacuum.

Ethyl formate

Ethyl formate formulated as an 11% (vol/vol) mixture with carbon dioxide has now been approved for use on grain in Australia but only by licensed fumigators. It is a naturally occurring compound found in a wide range of fruits, vegetables, cheese and grain products. Short exposure times of 3–6 h assist efficient application. Research has indicated its potential for use in a phytosanitary role against surface pests on grapes, bananas and other horticultural produce, against mites, thrips, leafrollers, aphids, mealybugs and spiders (Ryan and Bishop, 2003). In a brief summary, Krishna *et al.* (2005) report that ethyl formate plus carbon dioxide has potential as a phytosanitary treatment for bananas from the Philippines infested with various surface pests.

Propylene oxide

Propylene oxide at low pressure (13.3 kPa) provided mortality to four species of stored-product pests equivalent to methyl bromide (Isikber *et al.*, 2004). Propylene oxide is considered 'possibly carcinogenic to humans' and is banned in some countries (WHO, 2007). This determination is based on animal studies that provided 'sufficient evidence' of carcinogenicity.

Methyl iodide

Methyl iodide (iodomethane) is toxic to pests at comparable or lower rates than methyl bromide, but is a liquid at room temperature (boiling point 42°C). Soon after efforts to reduce methyl bromide use began, methyl iodide received attention as a possible replacement for postharvest uses. For example, Aung *et al.* (2004) concluded that methyl iodide at 26 g/m^3 for 2 h at 21°C followed by 24 h forced aeration to reduce toxicity to lemons, could provide an efficacious phytosanitary treatment against red scale, *Aonidiella aurantii*. The status of methyl iodide in some countries as a potential human carcinogen, based on a couple of studies with rodents that gave marginally positive results, has for the time being sidetracked the process for its approval on food, although registration is being sought for pre-planting uses. Because of lack of relevant data, the World Health Organization (WHO) has made no determination on possible carcinogenicity of methyl iodide to humans (WHO, 2007). Methyl iodide is used in Japan for phytosanitary treatment of timber.

Hydrogen cyanide

Phytosanitary uses of hydrogen cyanide (HCN), such as for treatment of cut flowers, dormant nursery stock and dried plant products, were largely replaced by methyl bromide. However, recent restrictions on the use of methyl bromide are resulting in the phytosanitary uses of hydrogen cyanide being considered again. Cut flowers and dormant nursery stock were often damaged by hydrogen cyanide in the past. Hansen *et al.* (1991a, b) observed that some Hawaiian cut flowers and foliage tolerated doses to kill some, but not all, quarantine pests found on these products. Summarizing on preliminary research, Park *et al.* (2006) concluded that hydrogen cyanide could be used to disinfest oranges of surface pests with insignificant injury to the fruit.

Methyl isothiocyanate

Methyl isothiocyanate has been identified as responsible for the insecticidal properties of the foliage of *Boscia senegalensis*, used in Niger as a traditional means of protecting stored cowpeas from insect attack. In the past it has been marketed commercially as a soil fumigant active against fungi, nematodes and insects although it is not clear whether any approvals are current. Ducom (1994) reported an in-depth study of the chemical as a potential grain fumigant, using the granary weevil, *Sitophilus granarius*, as the test insect. Because it sublimes at ambient temperatures it can be incorporated with a grain flow but if applied to a bulk would require continuous re-circulation. A related characteristic is its extreme susceptibility to sorption. Its potential as a replacement for methyl bromide is unclear at this time.

Cyanogen

Another member of the cyanate family of compounds, cyanogen (C_2N_2) is also known as ethanedinitrile. It is a long-known compound with wide potential as a soil and commodity fumigant and which also has systemic activity when used on living plants. The Commonwealth Scientific and Industrial Research Organization (CSIRO), Australia patented formulations as fumigants for protection of stored products, structures and timber (CSIRO Entomology, 2000). Wright *et al.* (2002) provided a short briefing at the 2002 Methyl Bromide Alternatives Outreach (MBAO) Conference on its characteristics, pending approval in Australia for use on timber and packing material as a potential replacement for methyl bromide. Its future as a broad-use fumigant for stored and fresh commodities is unclear at this stage.

Carbonyl sulphide

The use of carbonyl sulphide (COS) as a fumigant was patented by the Australian research organization CSIRO in 1992 and showed promise as a replacement for methyl bromide for use on grains and other stored products and durable agricultural commodities (Desmarchelier, 1994). Carbonyl sulphide is easily handled and applied; however, it requires constant circulation, is generally not as toxic to pests as methyl bromide, and leaves an objectionable odour that will dissipate after some time. For fresh commodities, Obenland *et al.* (1998) concluded that lemons would tolerate the > 8 h fumigation time required to kill Mediterranean fruit fly. An objectionable odour in the fruit found upon fumigation disappeared after 48 h. A dosage of 80 g/l resulted in only 87% codling moth mortality in nectarines (Aung *et al.*, 2001).

Superseded fumigants

The following fumigants were used in the past but are not currently used for phytosanitary reasons that we are aware of today. It is always possible that some may be resurrected in some formulation in the future.

Ethylene dibromide

Ethylene dibromide is no longer approved as a treatment for fresh fruits and vegetables in most countries. Historically, it was highly effective as a phytosanitary fumigation treatment against juveniles of tephritid fruit flies in fruit and vegetables. It replaced methyl bromide for many usages on fruits and vegetables for fruit flies from the 1950s because it was effective at lower doses and caused less phytotoxic injury. However, solanaceous fruits were intolerant, leading to use of pesticides (Swaine *et al.*, 1984a) or heat treatments (Sugimoto *et al.*, 1983; Sunagawa *et al.*, 1988) for these commodities.

Withdrawal of approvals for use of ethylene dibromide in the USA in 1984 (EPA, 1984) left a hiatus in phytosanitation only partly filled by methyl bromide. It was disapproved because it was considered a probable carcinogen following discovery of high levels in ground water owing to its use as a soil fumigant against pathogens and nematodes. However, it was also widely used at that time as an octane enhancer in gasoline and as an industrial solvent as well as numerous lesser uses. Other countries followed the USA, initially by reducing MRL and subsequently by delisting them. Few countries would currently permit its use in phytosanitation and it is effectively prohibited in world trade goods by the absence of MRL in major importing countries such as the USA.

Carbon tetrachloride

This fumigant was widely used for small bulks of grains and mill machinery until withdrawal of approvals for its use owing to its carcinogenicity. It has no known current approvals for use as a fumigant, but is still used in the manufacture of a number of industrial products where its carcinogenic properties can be managed.

Carbon disulphide

This is one of the oldest fumigants in use, dating from 1854 (Winks, 1984). Its main use has been against grain pests when it is sometimes mixed with carbon dioxide to minimize the risk of explosion, to which it is very vulnerable. It is applied as a liquid to a mat on the surface of a grain bulk or stack and allowed to evaporate to create the fumigant atmosphere, unless formulated with carbon dioxide when it is applied from a gas cylinder. Difficulties in application generally militate against its widespread use.

Conclusions

Fumigation should continue to be a major disinfestation methodology into the foreseeable future but it will become increasingly regulated with respect to workplace safety and food residues. The advantages of low cost and simple application technology make it suitable for use in less developed countries. It is also very appropriate as a first line defence in dealing with many outbreaks of invasive pests in new areas, especially those reliant on regulated pest free status.

New fumigants may emerge, but collection of data to acquire approvals is costly and no commercial organization can afford to bear the cost without the protection of a patent.

It can be expected that with time the proportion of phytosanitary pest management attributable to fumigants will be eroded by physical and management strategies, but that it will continue to be a significant proportion of overall phytosanitation treatment.

11 Disinfestation with Modified (Controlled) Atmosphere Storage

Atmospheres with low concentrations of oxygen (O_2) and/or high concentrations of carbon dioxide (CO_2) can achieve disinfestation to the degree required by phytosanitary treatments without unacceptable injury for some fresh commodities (Carpenter and Potter, 1994; Hallman, 1994a). Modified atmospheres for pest management of grain and other stored commodities, more tolerant to modified atmospheres than fresh commodities, are presented in Calderon and Barkai-Golan (1990). Modified atmospheres are not presently used as phytosanitary treatments but are used in some countries for pest management in stored commodities.

Dry atmosphere near sea level is composed of about 78% nitrogen (normally inert), 21% O_2, 0.93% argon (an inert noble gas), 0.04% CO_2 and trace amounts of other noble gases, methane and hydrogen. Of course there are thousands of other gases present in the atmosphere at any given time and place, some causing significant local reactions, but their amounts are miniscule on a global basis. Water vapour is present in variable amounts, usually making up 1–4% of the total. A modified atmosphere, the generic term, is a significant deviation from this atmospheric composition (Calderon, 1990). A controlled atmosphere is a modified atmosphere that is held at constant, known concentration tolerances by addition to or generation of desired levels of gases or purging excess quantities of gases. A more accurate term might be controlled modified atmosphere, as ambient atmospheric conditions can be controlled as they are in some buildings for the benefit of the habitants.

Low pressure is predominantly a low O_2 (modified atmosphere) treatment, with mode of action less related to the physical effect of low pressure or desiccation (Mbata and Phillips, 2001). Coating and bagging commodities may also kill pests inside by modifying the atmosphere in the commodity (Jang, 1990; Dentener *et al.*, 1992; Hallman, 1997).

Hallman (1994a) identified a number of controlled atmosphere phytosanitary treatments that seemed likely to work against several tortricid species, two mites, two thrips, an aphid, San Jose scale, *Quadraspidiotus perniciosus*, and sweetpotato weevil, *Cylas formicarius elegantulus*, on apple, asparagus, strawberry, sweet potato and walnut. Considerable additional research has been done since then. However, the only commercial modified atmosphere phytosanitary treatment remains one successful shipment of 225 kg asparagus from New Zealand to Japan that was disinfested of New Zealand flower thrips, *Thrips obscuratus*, and green peach aphid, *Myzus persicae* (Carpenter and Potter, 1994). The treatment consisted of exposing asparagus to 60% CO_2 in air, yielding about 8% O_2, for 4.5 days at 0–1°C. Although no live quarantine pests were found and asparagus quality was not significantly affected, Japan did not permit additional shipments, but the reason for this decision is unknown to us.

Efficacy of a modified atmosphere treatment at temperatures near 0°C may be due more to the low temperature than the modified atmosphere (Hallman, 1994a). Synergism between modified atmospheres and cold applied at the same time does not seem to occur. However, a 20 h exposure of mandarins infested with Mediterranean fruit fly, *Ceratitis capitata*, to 95% CO_2 at 25°C before cold treatment at 1.5°C in ambient atmosphere reduced lethal times to less than half of that required without the modified atmosphere pre-treatment (Alonso *et al.*, 2005). The 20 h pre-treatment by itself as well as the control had about 30% mortality; fruit were infested by placing ten diet-reared third instars 20 mm deep inside the pulp under a 10 mm-wide bored core.

Controlled atmosphere storage of fresh fruits is used to extend market availability beyond what is possible with normal cold storage (Thompson, 1998). It is a common storage regime for pome fruits using atmospheres of about one-tenth of the normal O_2 concentration and about 100 times the ambient CO_2 concentration (still only a few percent CO_2). Apples may be stored for up to 11 months using this system. At the temperatures (about 0.5–2°C) and times (at least a few months) regardless of atmospheric concentrations, disinfestation of quarantine pests would be accomplished.

One reason why modified atmosphere treatments have not been used commercially, save for the one instance in New Zealand, is that they are difficult to research and implement on a commercial scale. Of the commonly studied phytosanitary treatments, modified atmospheres may be the most difficult to study. For example, though Hallman (1994a) identified several possible treatments that might work commercially, he noted that further research was probably needed for all of them before they could be confidently used.

One of the complicating factors for modified atmosphere research is the response of insects to elevated CO_2 (Fleurat-Lessard, 1990). Hallman (1994a) cites several examples among a fruit fly, stored-product weevils, and tortricid moths where increased reaction to increasing CO_2 levels did not follow a typical dose-response curve. Green peach aphid response to increasing CO_2 concentrations was steeper at lower than at higher doses (Epenhuijsen *et al.*, 2002). The authors compared their results with others for grain weevils, crickets, thrips and other aphids and hypothesize that there is an active response to CO_2 in the 10–20% range and a slow response in the 20–60% CO_2 range.

Response of pests to O_2 levels may be no less complicated than response to CO_2 and results of combinations of modified levels of both gases can be exceedingly difficult to examine, model and predict (Fleurat-Lessard, 1990).

However, because the modes of actions of hypercarbia and hypoxia are different, the combination should be more effective than either modification alone. Another complicating factor is temperature, with increasing temperature usually increasing mortality.

Limitations to its applicability include product injury tolerance over the exposure times in the constituent atmospheres necessary and the overall logistical suitability. For fresh horticultural produce, a modified atmosphere is normally combined with low temperatures, which would contribute to pest mortality. For durable commodities, such as grain and other stored products, relatively high ambient temperatures are possible and these will usually accelerate pest mortality. An advantage for modified atmosphere disinfestation is the absence of residues that characterize chemical fumigants and other pesticides. However, a major safety hazard exists for operational staff from the anoxic and hypercarbic nature of modified atmospheres, which could asphyxiate unprotected operators.

Methods for Modifying Atmospheres

Storage atmospheres can be modified in the following ways:

- Synthesis of the storage atmosphere in which the space is purged of ambient air with nitrogen or otherwise displaced with a synthetic atmosphere of predetermined proportions of O_2, CO_2 and a biologically inert component, usually nitrogen. All three gases are available commercially in most countries. Use of other inert gases in this way, such as helium, is possible.
- Introduction of CO_2 to the air in the storage chamber, which would have the effect of reducing the proportions of ambient O_2 by a factor related to the amount of CO_2 added.
- Depletion of ambient O_2 and creation of CO_2 by passing ambient air through a 'burner' which consumes a flammable hydrocarbon gas by open flame, internal combustion or a catalytic process, resulting in an atmosphere of 1% O_2 and 12% CO_2, with the balance being made up of nitrogen and products of combustion (Banks *et al.*, 1990).
- Hermetic or semi-hermetic storage in which the respiration of the commodity and organisms infesting it deplete the level of O_2 and increase the level of CO_2. This can be induced in many ways ranging from ancient pit storages to modern packaging and coatings.
- Vacuum which lowers concentrations of all atmospheric gases equivalently. Vacuum is essentially a low O_2 treatment.

Maintenance of the storage atmosphere at the required gas composition levels depends on the gas loss from the structure. This is influenced by the permeability of the structural fabric and the effectiveness of seals at joins and entry points. Surface to volume ratio has a major influence (Fig. 11.1). Sheeted bulks are affected by wind which can induce a 'pumping' effect. Gas tightness of a structure or enclosure is readily calibrated and monitored by a water gauge in terms of pressure half-life. Ideally, any loss should be less than a proportionate 0.05/day. CO_2 levels can be reduced also through internal adsorption. With low

Fig. 11.1. An hermetically sealed grain storage with a capacity of 100,000 t (Moree, NSW, Australia) intended for modified atmosphere pest control. Structures of this size have low loss rates of the modified atmosphere because of their favourable surface to volume ratio.

O_2 systems it is almost inevitable that the atmosphere will need to be maintained in some way (Banks *et al.*, 1990). Modified atmosphere storage needs to be monitored and component levels supplemented as needed. Electronic equipment is available to monitor levels of individual components.

The technology of modification of structures for modified atmosphere storage of grain has been researched extensively (Banks *et al.*, 1990) and both horizontal and vertical sealed storages have been constructed in Australia. However, in practice it was found that these storages were utilized more for cost-effective fumigation with low doses of phosphine, which enabled access more readily to part contents within the storage term.

Mode of Action

Three characteristics of modified atmospheres, low O_2, increased CO_2 and temperature, contribute to pest disinfestation over a time span which may be as short as a few hours at warm temperatures to weeks at cool temperatures. There is considerable interaction between the components, sometimes being synergistic. The overall mode of action of modified atmosphere is not well known, nor, not surprisingly, is that of each of the individual components (Fleurat-Lessard, 1990). Reduced O_2 levels are toxic because O_2 is essential to vital functions of aerobic organisms. Critical levels can vary depending on the ability of an organism to accumulate glycolytic products, reduce its metabolic rate and restrict water loss (Carpenter and Potter, 1994). Enhanced CO_2 levels affect respiration including spiracular function of terrestrial arthropods and gas exchange across respiratory membranes. The effect of varying concentrations is highly complex as there is not a simple linear relationship between concentration and mortality, for example the presence of some O_2 compared to its virtual absence can accelerate mortality (Fleurat-Lessard, 1990).

Low O_2 levels resulting from any of the methods of atmosphere creation will be a major cause of insect death, although there will be a narcosis effect at high nitrogen concentrations which may delay death (Fleurat-Lessard, 1990). This is also the primary effect in hermetic storage without the introduction of gases (i.e. the CO_2 concentration resulting from reduction of pest or plant respiration is not high enough to have a significant effect) (Fleurat-Lessard, 1990). For the confused flour beetle, *Tribolium confusum*, the critical minimum level of O_2 was determined at 0.9% while for the rice weevil, *Sitophilus oryzae*, it was 0.15% (Fleurat-Lessard, 1990). This indicates the variation which may be present in the tolerances of pests, although the general level of O_2 for immediate mortality was thought to be around 1%. Considerable time may be required for mortality of pests. For example, for an atmosphere created with a fuel burner resulting in 0.5% O_2, 113 h of exposure at 18°C were required to achieve mortality of larvae of *Tribolium* spp. and at 1.2–1.5% O_2 at 14–15°C 3 months were required for complete disinfestation, although at that temperature the population would not increase (Fleurat-Lessard, 1990).

High CO_2 concentrations lead initially to a narcotic effect culminating in 'knockdown' (Fleurat-Lessard, 1990). This effect is useful as a laboratory handling technique especially for Lepidoptera as restoration to a normal atmosphere after a few minutes results in recovery with no apparent adverse effects. Longer exposure times are lethal so that above about 35% the LC_{99} can be as little as 10 days. Species can be concentration dependent, with the lethal response of some species, such as *Tribolium* spp., increasing throughout the range from 35 to 100% concentration. Other species are less concentration dependent. In the bean weevil, *Acanthoscelides obtectus*, variation in tolerance occurs and is presumably genetically controlled (Fleurat-Lessard, 1990).

Effect of Temperature on Modified Atmosphere Efficacy

The major factor affecting modified atmospheres, often as significant as the atmospheric modifications themselves, is temperature. In general, increasing the temperature increases efficacy of modified atmospheres. At low temperatures the lethal effect may be due more to the cold temperature than the modified atmosphere. For example, Mitcham *et al.* (1997) achieved higher mortality of a thrips and a tortricid under controlled atmospheres at 0°C compared with 5°C and hypothesized that a possible increase in metabolism at the lower temperature resulting from the insects preparing cryoprotection may have resulted in a greater efficacy of the controlled atmosphere. Examination of the small graphs giving the raw data in that publication seems to show that mortality in air was generally also greater at 0°C than 5°C for these two insects, which is to be expected, thus demonstrating that the colder temperature by itself had a significant effect on mortality. What is surprising is that a mite included in this study generally showed greater mortality in both modified atmosphere and air treatments at 5°C compared with 0°C. That result does not follow convention.

Soderstrom *et al.* (1991) recognized that low temperatures were causing significant mortality of codling moth, *Cydia pomonella*, when they observed higher mortality at 0°C than 5°C modified atmosphere treatments. In this case raw data were not given; it is useful to present raw data in graph or table form in publications for future comparisons, such as we would like to do now.

The relationship between modified atmospheres and temperatures near 0°C is not clear: perhaps there is some synergy based on some physiological reason as Mitcham *et al.* (1997) hypothesize or perhaps the relationship is basically additive, with cold providing significant mortality. At cool temperatures that are not generally lethal to insects but low enough to inhibit development, the cool temperature seems to be antagonistic to the efficacy of modified atmospheres (Soderstrom *et al.*, 1991). This seems logical in that insects that are metabolizing at a very low rate may be less susceptible to deficiencies or excesses. Knowing where the threshold for antagonism ends above the cool temperature range and where mortality increases again below it (whether due to simply the cold temperature or some interaction of cold temperature and modified atmosphere) is the key to optimizing modified atmosphere phytosanitary treatments.

Modified Atmosphere–Heat Synergy

Although the relationship between modified atmosphere and temperatures near 0°C has not been adequately elicited, at the other end of the phytosanitary temperature treatment range, heat, it is quite well understood (in efficacy if not in physiology) and has been taken advantage of extensively in research. Treatments have been developed that might be ready for commercial implementation.

In in vitro studies, Whiting *et al.* (1992a, b) developed a potential treatment consisting of 0.4% O_2 and 5% CO_2 at 40°C and requiring an estimated 4.2 h to achieve 100% mortality of four tortricids. At 30 and 20°C, respectively, the LT_{99} for fifth instar *Epiphyas postvittana* was about 8 and 22 times what it was at 40°C (Whiting *et al.*, 1991). Large-scale confirmatory testing is needed to set the minimum treatment time duration before this treatment could be recommended for commercial use. The most tolerant stage was the fifth instar and the most tolerant of the four species was the codling moth. Non-diapausing fifth instar codling moths were slightly more tolerant than diapausing ones.

When Whiting *et al.* (1995) exposed eggs and first, third and fifth instars of six leafrollers (Tortricidae) to three different atmospheres (1.2% O_2 and 5% CO_2; 4.2% O_2 and 5% CO_2; and air) at 40°C, fifth instars were not always the most tolerant (Table 11.1). However, the most tolerant of the six species was *E. postvittana*, and the fifth instar was more tolerant than any stage tested of any of the six species. Thus, a treatment for fifth instar *E. postvittana* should control all stages of concern of the other species as well, although one could raise a question about second and fourth instars, which were not tested.

Table 11.1. Estimated lethal times (LT_{99}) for most tolerant of 3 day-old eggs and first, third and fifth instars of six leafrollers (Tortricidae) exposed to three atmospheres at 40°C (Source: Whiting *et al.*, 1995).

	Most tolerant stage and LT_{99} (h) per atmosphere		
Species	1.2% O_2, 5% CO_2	4.2% O_2, 5% CO_2	Air
Ctenopseustis obliquana	3rd, 3.3	5th, 3.8	3rd, 5.6
Ctenopseustis herana	1st, 3.9	All stages, ~ 3.7	3rd, 6.1
Planotortrix excessana	All instars, ~3.1	3rd, 4	3rd, 5.6
Planotortrix octo	1st and 5th, ~4	3rd, 5	Egg, 5.8
Cnephasia jactatana	All instars, ~3.3	3rd and 5th, 4.3	All instars, ~ 5.8
Epiphyas postvittana	5th, 8.7	5th, 15.5	5th, 21

In the three previously cited in vitro studies (Whiting *et al.*, 1992a, b, 1995) the most tolerant stage apparently differed with O_2 level for some leafrollers. When the O_2 level was 0.4% the fifth instar of *Ctenopseustis obliquana* and *Planotortrix octo* was most tolerant; when the O_2 level was 1.2 or 4.2% or ambient, other stages were often more tolerant (Table 11.1).

Neven *et al.* (2006b) developed two treatments against codling moth and oriental fruit moth, *Grapholita molesta*, in peaches and nectarines for shipment to Pacific rim countries (codling moth) and Mexico and western Canada (oriental fruit moth). The treatments consist of 1% O_2, 15% CO_2, > 90% relative humidity (RH), air speed between 1.2–2.0 m/s, and heating times of either 12 or 24°C/h. The difference between the two treatments is speed; at the 12 or 24°C/h heating times, respectively, the treatments can be accomplished in about 3 and 2.5 h. No differences in fruit quality were observed between the two heating times.

The codling moth was determined to be the more tolerant of the two tortricids and large-scale testing was done with that species. The fourth instar was chosen as the most tolerant stage for both species, although it seems that it was the least tolerant among the larvae for both based on the LT_{99} estimates (especially upper 95% confidence limits; CL) and at least the raw data (raw data are in graph form and a little hard to decipher) for the codling moth (Table 11.2). For example, the estimated upper 95% CL of the LT_{99} for the codling moth third instar was over 1 h greater than the estimate for fourth instar.

Neven *et al.* (2006b) used the LT_{50} to determine tolerance; we suggest that the LT_{99} would have been more appropriate. Phytosanitary treatments must be efficacious at high levels of control (near 100%), so the dose that yields at least 99% control, the highest dose that statistical analyses confidently give, should be used to determine the most tolerant stage. Using lower levels of control to determine the most tolerant stage would be valid only if responses for the stages under question were parallel.

Nevertheless, the authors may still be on the right track. The concept of most tolerant stage should base determinations on the same objective (see Chapter 6). In this case larvae were determined to be dead if they showed no movement for up to 7 days after treatment. First to third instars, although seemingly more tolerant than fourth instars, had more growth to undergo than the latter before they could

Table 11.2. Estimated lethal times via probit analysis of controlled atmosphere/heat treatment of instars of codling moth (CM) and oriental fruit moth (OFM) (Source: after Neven *et al.*, 2006b).

Moth and instar	Lethal times (h)[a]		
	LT_{50} (95% CL)	LT_{90} (95% CL)	LT_{99} (95% CL)
CM 1	1.99 (1.59–2.34)	2.71 (2.40–3.46)	3.19 (2.78–4.24)
CM 2	1.92 (1.36–2.33)	2.72 (2.36–3.64)	3.23 (2.77–4.54)
CM 3	1.04 (0–1.36)	2.49 (2.12–3.44)	3.22 (2.71–4.79)
CM 4	2.24 (2.08–2.41)	2.77 (2.59–3.05)	3.14 (2.90–3.52)
CM 5	2.01 (1.74–2.26)	2.61 (2.37–3.06)	3.01 (2.70–3.63)
OFM 1	1.96 (1.64–2.07)	2.67 (2.47–3.01)	3.13 (2.86–3.60)
OFM 2	1.89 (1.67–2.10)	2.68 (2.46–3.02)	3.18 (2.89–3.64)
OFM 3	2.15 (1.88–2.58)	2.60 (2.29–3.19)	2.91 (2.55–3.63)
OFM 4	2.02 (1.84–2.21)	2.50 (2.31–2.82)	2.82 (2.59–3.27)

[a] The analyses were all significant at the 0.0014 probability level or better for time and for *y*-intercept except for CM 3.

complete development. Still, if one were to use this argument to determine the most tolerant stage, it would be necessary to follow continued development of early instars that survived treatment as well as convince regulatory agencies that live early instars found during inspection are of no concern because they will die later. This argument has been successful for irradiation phytosanitary treatments (Chapter 9) that generally do not cause acute mortality.

Tolerance of egg stages (whitehead, red ring and blackhead, in ascending order of development) was based on eclosion. Estimates of the upper 95% CL of the LT_{99} was 6.92 h for the red ring stage of oriental fruit moth, which is almost twice the maximum upper 95% CL for any of the instars of oriental fruit moth. However, as argued before with early instars, the organism at the egg stage has even more growth to do before the organism completes development. Nevertheless, because eggs and instars earlier than the fourth were estimated to require longer times to achieve high levels of mortality for both species, it would be prudent for the researchers to do some confirmatory testing. The apparently most tolerant stages, such as red ring egg and third instar, should be tested at the times found to suffice for fourth instars and the development of any survivors followed before recommending this treatment commercially.

Neven and Rehfield-Ray (2006) conducted a similar study for codling moth and oriental fruit moth on apples with similar questions resulting. The authors again chose the fourth instar as the most tolerant although analyses (raw data not given) indicate that the fifth instar might be the most tolerant for codling moth and all other oriental fruit moth stages seem slightly more tolerant than the fourth when based on the LT_{99}. Also, stopping egg hatch required longer times, over 1.5 times as much for whitehead oriental fruit moth egg compared, for instance, with the fourth instar. Again, it would be prudent to do some confirmatory testing with those stages estimated to have higher LT_{99} (upper 95% CL) values than fourth instars before recommending this treatment to industry.

Heating rate may affect the efficacy of heat/modified atmosphere treatments as it does heat treatments (Chapter 8). As the time to bring the chamber up to operating temperature increased in a treatment consisting of 1% O_2 and 1% CO_2 at 40°C, the estimated time required for the treatment to achieve 99% mortality decreased, until it passed 7.5 h after which it increased again (Fig. 11.2; Whiting and Hoy, 1998). The hours required to complete treatment at a heat-up time of 12 h was roughly equivalent to one of 6 h, showing that temperature adaptation by the pest had occurred. Under this treatment scenario, barring any differential effect on the commodity being treated, the best treatment would be the one that achieved quarantine security in the shortest amount of total time (heat-up plus treatment time), given that the modified atmosphere was being maintained during heat-up (i.e. the treatment with the shortest heat-up time).

Coatings of fresh commodities designed to prolong shelf life may modify the atmosphere inside and have also been shown to synergize heat treatments. Hallman *et al.* (1994) reduced by half the heated-air treatment time required to kill Caribbean fruit fly, *Anastrepha suspensa*, in grapefruits by coating the fruit before heating.

Hypobaric Storage as a Modified Atmosphere Treatment

Low pressure (hypobaric) storage, used to prolong shelf life of fresh commodities (Burg, 2004), may achieve control of pests, and the mode of action is considered to be due to reductions in O_2 levels with insignificant physical effects of low pressure per se or dehydration (Mbata and Phillips, 2001). The advantage of a hypobaric treatment is that gases need not be created, introduced or even monitored. The key variable is maintenance of the desired low pressure.

Hypobaric storage may be an ideal disinfestation technique for some difficult-to-treat fresh commodities, such as lettuce. Partial vacuum is already used to cool lettuce after harvest. Complete control of two aphids was achieved in 4 days at 5°C using a vacuum to initially remove air followed by insertion of 6% CO_2,

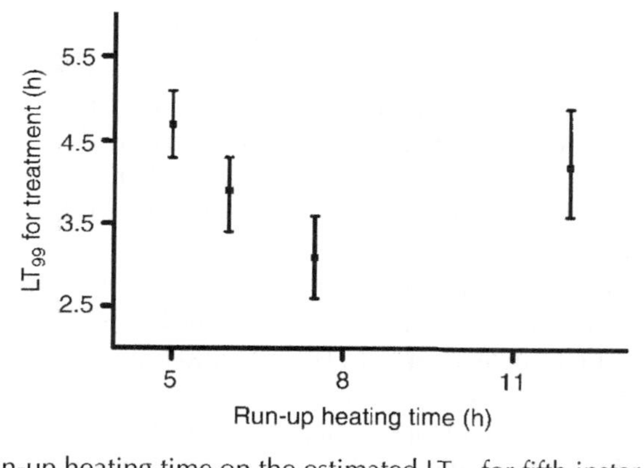

Fig. 11.2. Effect of run-up heating time on the estimated LT_{99} for fifth instar *Epiphyas postvittana* subjected to 1% O_2 and 1% CO_2 at 40°C (Source: after Whiting and Hoy, 1998). Run-up heating time is the time interval to bring the treatment chamber up to the operating temperature of 40°C from a beginning temperature of 20°C.

although the treatment achieved only 95% control of larvae of the leafminer, *Liriomyza langei* (Liu, 2003). Previous research with lettuce has usually found the commodity to be intolerant of even small amounts of CO_2. The technique should also be investigated for cut flowers where few viable disinfestation alternatives to fumigation have been found.

All Caribbean fruit fly eggs and larvae in agar diet were killed in about 9 days upon exposure to a vacuum of 15 mm Hg at 13°C (Davenport *et al.*, 2006). In small-scale tests, mangoes, carambolas and guavas survived the treatment well.

Target Pest Groups

Different pest groups may respond differently to modified atmospheres. More progress has been made and more conclusions about pest response can be drawn from stored-product pests (Fleurat-Lessard, 1990; Mbata *et al.*, 2004). However, responses of stored-product pests may differ from pests infesting fresh commodities because pests on fresh commodities are generally not subject to low RH as are pests on stored products. Also, durable commodities tolerate more extreme treatments than fresh ones.

Fruit flies

Fruit flies of the family Tephritidae are the most important group of quarantine pests across the spectrum of fresh fruit traded internationally. However, fruit flies have not received as much attention from modified atmosphere researchers as their importance might indicate.

Benschoter *et al.* (1981) seem to be the first to have tried this technique against a tephritid, the Caribbean fruit fly; 5-day-old larvae died after 60 h exposure at 22–23°C in 100% nitrogen in vitro. Benschoter (1987) further assessed the response of Caribbean fruit fly eggs and larvae in vitro to modified atmospheres of 20, 50 or 80% CO_2 and 2, 10 or 20% O_2 (the balance made up of nitrogen) at 10 and 15.6°C. Increased mortality generally coincided with the highest CO_2 concentration regardless of O_2 concentration. At the lowest CO_2 concentration, however, increased mortality coincided with lower O_2 level. Mortality was somewhat increased at the higher of the two temperatures. Complete mortality of 150 insects tested occurred in 7 days with some of the treatment combinations (Table 11.3).

Prange and Lidster (1992) achieved at most 90% mortality of blueberry maggot, *Rhagoletis mendax*, after 48 h in various levels of CO_2 at 21°C (Fig. 11.3). Mortality reached a peak at about 70% CO_2 and then declined until less mortality was achieved at 100% CO_2 than at 50% CO_2.

Complete mortality of apple maggot, *Rhagoletis pomonella*, larvae in apples was achieved within 14 days at 10°C in atmospheres with 15 or 19% CO_2 (the balance nitrogen) (Agnello *et al.*, 2002). Under the same conditions at least 3% of eggs survived to produce third instars after removal from the treatment conditions.

Table 11.3. Modified atmosphere combinations that provided 100% mortality of Caribbean fruit fly, *Anastrepha suspensa*, eggs and larvae in vitro (n = 150). All combinations of the three O_2 and CO_2 levels and two temperatures were tested (Source: Benschoter, 1987).

O_2 (%)	CO_2 (%)	Temperature (°C)	Time[a] (days)
2	20	15.6	7
2	20	15.6	10
2	50	15.6	7
2	50	15.6	10
2	50	10	10
2	80	15.6	7
2	80	15.6	10
10	20	15.6	7[b]
10	20	10	10
10	50	15.6	7
10	50	15.6	10
10	50	10	10
20	50	15.6	10
20	50	10	10
20	80	15.6	7
20	80	15.6	10
20	80	10	10

[a] Times were 3, 5, 7 and 10 days.
[b] This combination at a longer time (10 days) resulted in 98.6% mortality.

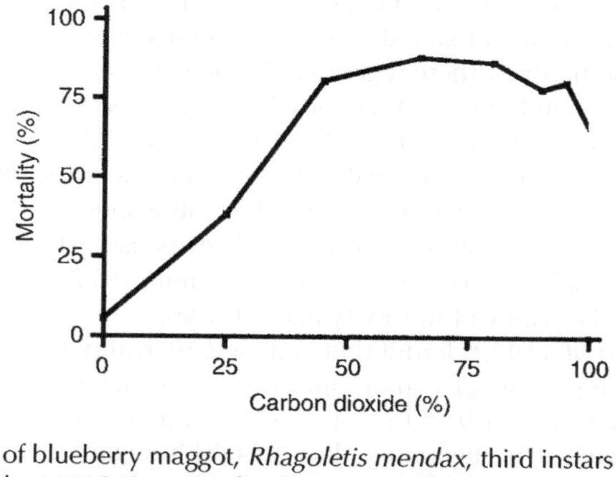

Fig. 11.3. Mortality of blueberry maggot, *Rhagoletis mendax*, third instars subjected to differing levels of CO_2 for 48 h at 21°C (Source: after Prange and Lidster, 1992). O_2 levels were either 2 or 5% (except for 100% CO_2) and data are means of both O_2 levels because authors found no difference between them.

Carpenter and Potter (1994) found that for tephritid fruit flies an atmosphere of 3% O_2 and 0–100% CO_2 at 0–20°C gave complete control of eggs and young larvae after 10 days or more of storage.

Fruit coatings provide some mortality of tephritid fruit fly immatures in fruit and the mode of action is probably via atmospheric modification inside the fruit (Hallman *et al.*, 1994; Hallman, 1997). The technique, already practised in many cases to preserve fruit quality (Krochta *et al.*, 1994) and not efficacious enough as a stand-alone treatment, might be incorporated as part of a phytosanitary system to reduce risk of infestation (Chapter 5). The added benefit to a phytosanitary system may require no additional action where coatings are already used except to document and quantify any risk reduction imparted by the coating. Coating is used as part of a physical phytosanitary treatment against a surface pest (mite) and that use is covered in Chapter 13.

Modified atmospheric packaging was studied as a phytosanitary treatment against fruit flies over a decade ago, but seemed too susceptible to variation in result and interruption of seal to be reliably efficacious to the degree of security required of phytosanitary treatments (Hallman, 1994a). None the less, modified atmospheric packaging might form part of a phytosanitary system in a like manner as coatings.

Lepidoptera

The largest group of pests studied with modified atmosphere phytosanitary treatments has been Lepidoptera, especially larvae of the fruit-boring family Tortricidae.

Commodities such as fruits of the family *Rosaceae* that are hosts to moth pests especially Tortricidae customarily tolerate storage under modified atmosphere for many months. Codling moth, *C. pomonella*, is the most important quarantine pest of these fruits internationally. It is this group of Lepidoptera pests of fruits against which modified atmosphere may have the greatest potential as a disinfestation treatment. The quality of stored pome fruits is preserved by modified atmosphere regimes and lethality of these regimes to lepidopterous pests is relevant. Toba and Moffitt (1991) found that mortality of all non-diapausing larvae of codling moth occurred in less than 13 weeks at 0.8–1.6% CO_2 and 2.2–3% O_2 at 0°C, which would be a useful minimum modified atmosphere storage time for quarantine purposes. These are the commercial modified atmosphere storage conditions for conserving apple quality for months. With diapausing larvae, 19 weeks were required for complete mortality (Moffitt and Albano, 1972). Low temperature may have been the key cause of mortality in both cases.

Gaunce *et al.* (1982) found that complete mortality of both diapausing and non-diapausing larvae of codling moth might be achieved after 33–48 h in a modified atmosphere of 95% CO_2 at 27°C, but injury to some apples was evident. This is analogous to fumigation with CO_2 as the fumigant. Experimental numbers were low and no statistical efficacy can be ascribed.

Codling moth is a species of quarantine importance in stored products of the families *Rosaceae* (almonds) and *Juglandaceae* (walnuts). Nuts are generally more tolerant of modified atmospheres and high temperatures than are fresh commodities. Soderstrom *et al.* (1996) tested atmospheres of 0.5% O_2 in nitrogen, 0.5% O_2 plus 10% CO_2 in air or 98% CO_2 in air against stages of codling moth infesting stored walnuts at temperatures of 39–45°C. They found that the 98% CO_2 atmosphere was the most efficacious. Higher temperatures were associated with more rapid mortalities. In testing, they used diapausing larvae because they were the most tolerant stage to modified atmospheres.

Other tortricid pests include the lightbrown apple moth, *E. postvittana*, and a number of other leafrollers native to New Zealand. These leafrollers cause surface damage to fruit as well as attacking foliage. Whiting *et al.* (1991, 1995) found that at temperatures $\geq$ 20°C, atmospheres low in O_2 and with moderate CO_2 levels would achieve disinfestation of fruit, increasing in efficacy at higher temperatures of up to 40°C. Interactions between levels of the modified atmosphere components and temperature were complex, but usefulness of modified atmosphere as a disinfestation treatment would depend on the compatibility of the raised temperatures with fruit quality and general postharvest handling practices. Multiple species of moth pests of stored products, especially grain, are commercially targeted with modified atmosphere as a pest management technology.

Hemiptera

Scale insects and mealybugs are the main groups of hemipterans likely to require phytosanitary treatments. Other families in this insect order that are often quarantine pests are aphids (Aphididae), whiteflies (Aleyrodidae), seed bugs (Lygaeidae) and leafhoppers (Cicadellidae).

Hard scales (Diaspididae) may be found on the surface of fruits but mealybugs (Pseudococcidae) and soft scales (Coccidae) may occur around the stem area of pome fruits and the blossom end of navel oranges. Hard scales are judged to be difficult to kill with modified atmosphere (Carpenter and Potter, 1994). Gaunce *et al.* (1982) recorded complete mortality of San Jose scale, *Quadraspidiotus perniciosus*, a phytosanitary pest of apples and other pome fruits, in a modified atmosphere of < 1% O_2 and > 90% CO_2 at 12°C after 2 days but only 50% mortality at the commercial storage temperature of 1°C after 5 days.

Contrary to most findings on the relationship between modified atmosphere and temperature, mortality of the longtailed mealybug, *Pseudococcus longispinosus*, decreased with an increase in temperature from 0 to 20°C in atmospheres containing 0, 9 or 18% CO_2 plus 2% O_2 with the balance nitrogen (Potter *et al.*, 1990). Complete mortality (*n* not given) after 2 weeks was achieved only at 0°C, 18% CO_2, while 18% CO_2 at 20°C gave 52% mortality. This result may be due to cold overriding the effect of the atmosphere. Under the same conditions the New Zealand wheat bug, *Nysius huttoni*, was easier to kill. Complete mortality after 2 weeks was achieved with 0°C, 18% CO_2 and 20°C, 9 and 18% CO_2.

Disinfestation of the obscure mealybug, *Pseudococcus affinis*, from apples could be achieved in a number of hours combining low O_2 with heat (Whiting and Hoy, 1997). As the treatment temperature approached 45°C differences in O_2 concentration had less effect on time needed. For example, at 40°C the time to achieve 99% kill of adult females at 5% O_2 (38 h) was almost three times that required at 1% O_2 (14 h) while at 45°C there was essentially no difference in time required (6 h).

Carpenter (1995) investigated several modified atmosphere/temperature combinations against green peach aphid, *M. persicae*, on asparagus and found that essentially a cold treatment alone (0–2°C in air for 4 days) was best for aphid kill and asparagus quality.

Thysanoptera

Thrips are becoming increasingly recognized as a pest group of major quarantine importance on fresh plant commodities traded nationally and internationally. Not only do they result in downgrading of the appearance of cut flowers and fruits but they are also vectors of potentially devastating viral diseases of plants. One of the most important thrips in trade in cut flowers and some fruits is the western flower thrips, *Frankliniella occidentalis*. Controlled atmospheres of low O_2 and high CO_2 have been used with some success against this species on strawberry fruits with risk of some off-flavours (Aharoni *et al.*, 1981). In fact, an atmosphere that might work to disinfest strawberries is 2 days with 2% O_2 and 90% CO_2 at 2.5°C.

Potter *et al.* (1994) found various combinations of CO_2, temperature and time (O_2 was fixed at 2%) that provided complete mortality (of a total of 300 thrips per treatment combination) of *T. obscuratus* adults. Among all combinations of five temperatures (0–20°C) and three CO_2 levels (0–18%) used, the following achieved compete mortality: 15°C, 18% CO_2 in 4 days; 10–20°C, 18% CO_2 in 6 days; 0°C or 20°C, 9% CO_2 in 6 days; all temperatures at 18% CO_2 in 8 days; and 0–5°C and 20°C, 9% CO_2 in 8 days. Large-scale testing would be required before any of these combinations could be used commercially.

Page *et al.* (2002) achieved 100% control of *Thrips tabaci* in onions at CO_2 levels $\geqslant$ 30% (balance air) after 24 h at 20°C. However, the control suffered high mortality, reaching > 70% in 12 h, casting doubt on the efficacy of the modified atmosphere.

Mites

Mites are said to be more difficult to kill with modified atmosphere than insects even though their activity stops at high concentrations of CO_2 (Fleurat-Lessard, 1990). Gaunce *et al.* (1982) recorded increased mortality of European red mite, *Panonychus ulmi*, and McDaniel spider mite, *Tetranychus mcdanieli*, on apples stored in modified atmosphere as compared with air cold storage over periods of 2–5 months. However, mortality was not complete, with survival ranging from 0.4 to 1.2%, respectively. Navarro *et al.* (1985) tested atmospheres of 2–21% O_2 and 10–40% CO_2 at 15 and 26°C against the grain mite, *Acarus siro*. Complete mortality required 3 days for 2% O_2 at 15°C and 5 days for 10% O_2 at 26°C. Complete mortality required 3 days at 26°C in 20% CO_2 and 4 days at 15°C in 30% CO_2. These results are indicative of the interactions which can occur between temperature, O_2 levels and CO_2 concentrations.

Whiting and van den Heuvel (1995) studied modified atmospheres against diapausing two-spotted spider mites, *Tetranychus urticae*. Reducing the O_2 concentration from 1.3 to 0.4% halved the LT_{99}; increasing the CO_2 concentration from 5 to 20% had a roughly similar effect. Temperature had a great effect on mortality, reducing the LT_{99} from 112 to 15 h as the temperature was increased from 20 to 40°C.

Commodity Quality

Modified atmospheres were used to preserve fresh commodity quality long before they were studied as potential phytosanitary treatments. However, the O_2 and CO_2 levels and temperatures used for phytosanitary purposes often differ from those used for commodity quality purposes in ways that may be detrimental to commodity quality. For example, most pome and stone fruits do not tolerate < 2% O_2 or > 5% CO_2 (Kader and Ke, 1994). Phytosanitary modified atmospheres ideal for pest mortality often contain < 2% O_2 and/or > 5% CO_2. Symptoms of low O_2 or high CO_2 injury to fresh commodities include internal and/or external browning, peel pitting, failure to ripen properly and increased rate of decay. Nevertheless, some insecticidal atmospheres may provide the beneficial effects of modified atmosphere storage as well as kill quarantine pests.

Because low O_2 results in anaerobic respiration, the products of this, such as ethanol and acetaldehyde, may yield detectable off-flavours (Kader and Ke, 1994; Shellie *et al.*, 1997; Alonso *et al.*, 2002). Off-flavours may dissipate with time.

Conclusions

Modified atmospheres have been shown to have the potential to be effective as a phytosanitary treatment against pests of fresh fruits, vegetables and cut flowers. The major disadvantage for its use on fresh commodities are the extended times required at low temperatures for complete pest mortality and the complicated nature of researching the different factors that affect efficacy. Long treatment times can be alleviated by using higher temperatures, but that may have consequences for commodity quality. At temperatures near 0°C it may be the temperature more than the modified atmosphere that is causing mortality.

Recommendations for Further Research

Researching modified atmospheres as phytosanitary treatments is complicated. However, the promise of developing efficacious treatments, especially at warmer temperatures, is worth the effort. To sort out the most promising combinations of O_2, CO_2, temperature and possibly other factors requires adequate experimental design including proper controls.

Low-pressure treatments, which function as modified atmospheres, deserve further research for a variety of fresh and stored products.

12 Postharvest Phytosanitary Pesticide Treatments

The scope for non-gaseous pesticide treatments against pests of phytosanitary importance is broad, ranging from field application in phytosanitary systems (Chapter 5) to postharvest disinfestation treatments. This is because pesticides are economical and easy to apply and arguably are the only major treatment that can offer residual protection. Residues are the major disadvantage of pesticides; legal safe maximum residue limits (MRL) are in place but there is often a negative perception of eating products with pesticide residues, however negligible. Pesticide label requirements are legally enforceable in most countries. They are based on toxicological studies and good agricultural practice and are recognized to result in no problems for the user or consumer if followed correctly.

This chapter concentrates on phytosanitary treatments using pesticides but gives examples where postharvest pesticides are used in phytosanitary systems. Most countries differentiate between field and postharvest application of pesticides and do not allow postharvest use except for certain fungicides, which are generally not for quarantine purposes in any case. The use of postharvest pesticide treatments on fresh commodities is becoming more and more restricted, regardless of whether MRL are met.

A comprehensive review of postharvest phytosanitary insecticide treatments for fruits and vegetables was done by Heather (1994). Snelson (1987) reviewed insecticides used on grains, seeds and other stored products in considerable detail. Usages and resultant residues are governed by regulatory approvals that vary among countries. Practical control of pesticide use on crops is accomplished by setting MRL and periodic analysis of commodity samples. MRL values may be searched online (FAO/WHO, 2007; FAS, 2007).

Postharvest, pesticides are used in phytosanitary roles to disinfest grains, seeds, fruit and cut flowers. The main postharvest usage of pesticides on fruits and vegetables for quarantine purposes has been by Australia and New Zealand as disinfestation treatments for tephritid fruit flies. Treatments can be as dips or flood-sprays applied to fruits and vegetables as they are processed for packaging

(Fig. 12.1). For cut flowers, which currently have fewer residue constraints than edible commodities, their variety and use is much more widespread. In the past, pesticides have been used in many countries admixed with harvested grains as disinfestants and protectants, but they are currently being reduced in this role as the marketing advantages of freedom from pesticide residues take effect. However, there is still considerable potential for postharvest phytosanitary use of pesticides, especially on cut flowers and seeds.

Safety Standards

All pesticide usage must meet relevant health and safety standards with respect to application and residues. Safety of operators, the public and consumers in the application of pesticides is subject to regulatory control by health and agricultural authorities according to the state or country. Establishing legal residue limits in food requires the coordination of toxicological studies, used to set an acceptable daily intake (ADI) for the pesticide by humans, and residue studies to provide data to set the MRL. These are often linked to withholding times to ensure that the pesticide will decay to acceptable levels before marketing of the treated commodity. Labelled instructions on purchased pesticide must be followed as the overriding authority on use.

The United Nations Food and Agriculture Organization, World Health Organization (FAO/WHO) Codes, Codex Alimentarius Commission, established a mechanism for governments to agree on MRL thus facilitating world trade in food

Fig. 12.1. A re-circulatory pesticide spray treatment module of a tomato packing line. The spray bars are seen on the top centre of the unit, the further set high pressure low volume and the nearer set high volume low pressure; the tank of insecticide can be seen under the unit. Application rates are controlled by pesticide concentration, nozzle aperture and conveyor speed.

commodities. Countries have either accepted FAO/WHO estimates of ADI and MRL or established their own. Use of a pesticide is permitted only if residues from that use are compatible with the ADI (Hamilton, 1988).

Criteria for Effectiveness

For fruit and cut flowers, postharvest phytosanitary treatments must achieve efficacy standards, which relate to recognized levels of quarantine security against pests, should be practical to handle and apply and not adversely affect the quality of a commodity. Before the treatment can be approved, it must be demonstrated to not adversely affect human health or the environment as used.

Postharvest phytosanitary treatments may be mandatory or discretionary (Chapter 6). Discretionary treatments against quarantine and regulated non-quarantine pests are prophylactic in nature and intended to ensure that a commodity is not rejected by an importing country following inspection on arrival. They may also be applied to meet export standards of a producer country. In this instance they can be in response to interceptions of categorized pests in inspections prior to export, or as a purely prophylactic measure especially against pest populations that might become apparent after extended transit times.

Insecticidal dips and packing-line flood sprays for fruit have been demonstrated to achieve a 99.99% level of efficacy against fruit flies in thin-skinned fruits such as tomatoes (Swaine *et al.*, 1984a; Heather *et al.*, 1987). On thicker-skinned fruits such as mangoes, the efficacy may be lower (Swaine *et al.*, 1984b). There are no known treatments with insecticide dips or sprays with a clearly demonstrated efficacy level of 99.9968% ('probit 9') in commercial use against fruit flies although this level of quarantine security would be equalled or exceeded where low initial infestation levels are achieved by efficient pest management during production or an intrinsically low incidence of infestation.

Saul and Seifert (1990) indicate 'probit 9' capabilities for the insect growth regulator methoprene, although this may have been due in part to the wax coating in which the methoprene was applied which may have resulted in a insecticidal modified atmosphere inside the fruit (see Chapter 11).

For commodities going to markets where no mandatory treatment is required it is usual that the product pass at least an import inspection. The sample size and frequency of this inspection governs the efficacy required of any pre-export pest management including disinfestation measures. This approach to quarantine security is frequently applicable to cut flowers and durable commodities such as grains and other stored products. It is achievable with pesticides, subject to conformity with MRL. For grains it is particularly appropriate as infestation levels of storage pests can increase by a factor of 50 during sea transit times in the absence of any residual effect of a disinfestation or protectant treatment.

For tomatoes shipped from Australia to New Zealand a maximum estimated infestation level of five fruit flies in a million tomatoes was accomplished through a phytosanitary system that included a 1 min postharvest dip in dimethoate. This system required registration of growers and defined field inspection procedures and pesticide sprays while crops were being grown. It also detailed packing-house inspections, the postharvest treatment, and formal sampling rates by Australian and New Zealand quarantine service inspectors together with rejection levels for other pests (Anon., 1993).

Pesticides for Fruit and Vegetables

Pesticides for postharvest use on fruits and vegetables must satisfy the strictest safety and residue criteria of any use on plants. Few insecticides satisfy these demands. Examples from three pesticide classes follow.

Organophosphorus pesticides

The use of dimethoate (De Pietri-Tonelli and Barontini, 1957) and fenthion (Unterstenhofer, 1960), insecticides with significant systemic action against the eggs and larvae of tephritid fruit flies in fruit, created a new approach to field control of these pests (May, 1962). Both insecticides are currently used in Australia as postharvest disinfestation treatments for a range of fruits, but it is possible that their use on fruits with edible peel may be curtailed in the near future (APVMA, 2007). Their postharvest efficacy ranges from 99.5 to > 99.99% at the 95% confidence level (CL). A 1 min, 400 mg/l dimethoate dip disinfestation treatment against Queensland fruit fly, *Bactrocera tryoni*, was the first accepted for a range of fruits and vegetables (Anon., 1982). Braithwaite (1963) reported postharvest application trials in which bananas were dipped for 1 min in 500 mg/l of dimethoate or fenthion in water as disinfestation treatments against Queensland fruit fly. Resulting residues for dimethoate at all times were < 2 mg/l, the Australian MRL (National Health and Medical Research Council, 1988). Australian registration of fenthion for use on bananas did not proceed. A commercial treatment unit applying a spray of 300 mg/l of dimethoate was used for a number of years in Queensland. Saunders and Elder (1966) and Smith (1977) reported similar trials against banana fruit fly, *Bactrocera musae*. In Taiwan, Lee (1968) reported on the efficacy of three organophosphorus insecticides against Oriental fruit fly, *Bactrocera dorsalis*, with best results from a 1 min dip in fenitrothion. Fenitrothion is not known to have an actively penetrating mode of action.

Swaine *et al.* (1984a, b) tested dimethoate on tomatoes and mangoes. For tomatoes a dip for 1 min in 425 mg /l of dimethoate resulted in 99.997% and 100% mortality in > 30,000 eggs, 24 h old and > 30,000 *B. tryoni* larvae, 5 days old, respectively. A 3 min dip caused 100% mortality on both eggs and larvae with residues that averaged 0.58 mg/kg on the day of treatment. For mangoes, a 3 min dip in 500 mg/l of dimethoate caused 99.98 and 99.97% mortality in > 30,000 eggs, 24 h old and > 30,000 larvae, 5 days old, respectively. When the insecticide was mixed with 55°C water and benlate to control anthracnose disease, insect mortality was slightly lower compared with dips at ambient temperature (20–30°C). Resultant residues of dimethoate were always below the Australian MRL of 1 mg/kg for tomatoes and 2 mg/kg for mangoes. Adverse tastes were not detected in any dipped fruit of either tomatoes or mangoes, and fruit quality was not affected. Swaine *et al.* (1984a) observed that in tomatoes third instars were more tolerant than eggs, and individuals occasionally pupated. Adults did not normally emerge from these puparia and quarantine security was unaffected.

Efficacy levels > 99.99% were reported by Heather *et al.* (1987) against *B. tryoni* for high volume re-circulatory flood-spray treatments of 400 mg/l of dimethoate or fenthion on tomatoes. Wetting times were equivalent to a 1 min dip. Neither treatment resulted in residues on the day of treatment in excess of the Australian MRL for fenthion (2 mg/kg) or dimethoate (1 mg/kg), nor were there any adverse effects on taste. Subsequent trials with these insecticides on rockmelons and zucchinis against cucumber fly, *Bactrocera cucumis*, gave similar results, again with no adverse taste effects (Heather *et al.*, 1992).

Dip and flood-spray treatments are valid alternatives to fumigation, especially for fruits or vegetables prone to fumigant phytotoxicity at the concentrations required. The dimethoate dip for tomatoes was developed primarily because they were susceptible to damage when fumigated with ethylene dibromide, a fumigant withdrawn from use for public health reasons (Chapter 10). Insecticide treatments also have a major benefit in that they provide residual protection against subsequent infestation.

Dip or re-circulatory flood-spray systems could increase the risk of spreading inoculum of postharvest rots, although this has not proved to be a problem in practice. Three other risks to the reliability of insecticide dips or flood sprays in closed systems are perceived. The first is 'stripping' of the insecticide, which occurs when the active ingredient is selectively removed from the dip formulation by adherence to the fruit (a characteristic of some fungicides). Neither the dimethoate dip life study by Noble (1983) nor measurements on experimental spray vats before and after fruit treatment (Heather, unpublished data) showed loss of active ingredient from this cause. The second risk to insecticide concentrations can occur where fruits are washed, then dipped or sprayed while still wet. Over time, water is added to the system which dilutes the dip. The third risk, chemical decay of the active ingredient, can be predicted and compensated for by adding more insecticide periodically (Noble, 1983).

Field applications of dimethoate or fenthion may be appropriate to meet quarantine security. An example is quarantine requirements for interstate trade in Australia between Queensland, where fruit flies are endemic, and Victoria, which is at the southernmost limit of distribution and virtually free of fruit flies (Bateman, 1967). Here, quarantine security is met in some circumstances with a grower's declaration that a dimethoate or fenthion spray was applied before harvest. This technique has proved effective for those crops in which good spray coverage of each fruit is possible. It can fail where fruit fly populations are high and spray coverage is inadequate.

Hallman and Foos (1996) increased the level of mortality of Caribbean fruit fly, *Anastrepha suspensa*, in grapefruits with dimethoate by applying the insecticide via fruit coatings. Dimethoate residue levels in grapefruit pulp with one coating were as low as or lower than dimethoate applied in water, although the level of fruit fly control was significantly higher. In another coating, dimethoate residue levels increased.

Organochlorines

A 1 min dip in 560 mg/l of endosulfan in water enabled disinfestation of exported Australian pineapples of dried fruit beetles, *Carpophilus* spp. (Nitidulidae) (Beavis *et al.*, 1991). This discretionary treatment satisfied an export inspection standard of no infested fruit per 600 sampled. Residues measured on the day of treatment were 1.9 mg/kg and, 7 days later, 1.4 mg/kg. The Australian, Japanese, USA and EU MRL for endosulfan on pineapple are 2, 2, 1 and 0.05 mg/kg, respectively. MRL values are expected to reflect good agricultural practice while at all times being below public health risk levels.

Dicofol is an acaricide that has potential for use immediately preharvest or postharvest against mites infesting ornamental flowers, fruits and vegetables but is not currently approved for postharvest use. The MRL for dicofol on fruit is 5 and 3 mg/kg for Australia and New Zealand, respectively. The USA has set no MRL for dicofol on fruit, and the limit in the EU varies from 0.02 to 2 mg/kg according to fruit group. Tetranychid mites, although recognized as cosmopolitan, can be categorized as regulated non-quarantine pests requiring discretionary disinfestation treatments although 'probit 9' mortality level is rarely required. However, when set tolerance levels apply at inspection, such as the maximum of 20 infested fruit in a sample of 600 used in Australia and New Zealand protocols, preharvest or postharvest population suppression measures are frequently required.

Insect growth regulators

Methoprene is a juvenile hormone analogue that prevents insects from developing to the adult. Saul *et al.* (1985, 1987) reported that methoprene applied to fruits provided significant control of tephritid fruit fly immatures therein. The methoprene was imbedded in a wax that provided up to 89% control by itself when the results of the wax-only controls are examined. No methoprene-only control (without the wax) was used to separate the effect of the wax or measure any interaction.

Major problems with using insect growth regulators in phytosanitary treatments are that larvae will be alive upon inspection and they could be liberated from the fruit when cut open by consumers before they have accumulated efficacious amounts of the growth regulator and develop to adults capable of reproduction. Saul and Seifert (1990) found that the methoprene concentration in the inner layer of pulp of treated papayas was 0.1% of the level in the peel. The first problem (live insects upon inspection) also happens with irradiation (Chapter 9) and, to a lesser extent, organophosphorus dip treatments. Strict regulation of pest management programmes during production and packing can help overcome this problem to some extent by reducing the chance of finding detectable infestations in export produce, but when live insects are detected regulatory action may be taken. With irradiation, finding live quarantine pests upon inspection has become accepted because it has been demonstrated that the process prevents further development and/or reproduction. This approach would not be recommended for insect growth regulators because the process is not immediate as it is for irradiation but gradually occurs as the pest feeds and accumulates uncertain doses of the growth regulator.

The use of insect growth regulators is a novel approach to phytosanitation, although they have been used in many other areas of pest management for a couple of decades, and deserves further research. Primary among research efforts would be assuring that efficacious levels of the insect growth regulator reach all quarantine pests before they can escape from the quarantine system. Although insect growth regulators do not pose detectable risks to human health, they may leave readily detectable residue levels in fresh produce and would not be accepted by organic producers.

Pesticides for Stored Products

The range of pesticides available for stored products including seeds and grains is broader than that available for edible fresh produce because stored products usually undergo considerable processing and time before consumption. Processing usually lowers pesticide levels, although in a few cases it may concentrate them (NRC, 1993).

Organophosphorus chemicals

Those currently approved for use against beetle and moth pests of grains and seeds include malathion, dichlorvos, pirimiphos methyl, chlorpyrifos methyl and fenitrothion (White and Blackett, 1988). Malathion is rarely effective now due to the widespread incidence of resistance in most species of grain pests. Resistance to the other organophosphorus pesticides above is also present in many species but, overall, does not affect efficacy to the extent occurring with malathion. However, management of resistant pest populations in grains and other stored products requires availability of a range of pesticides with differing modes of action, the more important of which are discussed in a subsequent paragraph. Dichlorvos at 6 mg/kg is currently effective against all stored-product pest species except *Carpophilus* spp. Chlorpyrifos methyl is used against non-resistant populations at 5 mg/kg with alternatives pirimiphos methyl at 4 mg/kg and fenitrothion at 6 mg/kg. These last three are effective against *Carpophilus* spp. Higher concentrations are used as grain protectants where storage times of 3–9 months are intended, enabling decay to the MRL or lower at the time of utilization or export.

Other stored-product pesticides

Natural pyrethrum and synthetic pyrethroids are widely used on grains and seeds. They are effective against many pests alone or may be used in combination with an organophosphorus pesticide where resistance management is required. Synthetic pyrethroids approved for use on grains and seeds include bioresmethrin, phenothrin and cypermethrin. Pyrethrum and pyrethroids can be enhanced by the addition of a synergist, piperonyl butoxide, enabling a lower concentration to be used, with cost normally the deciding factor. Insect growth regulators such as methoprene are approved for use in this way and may be used in combination with an organophosphorus chemical where resistant species are involved. Carbaryl is approved for some stored-product usages but is not accepted for malting barley owing to a detectable off-flavour in the brewed beverage.

Chemically inert desiccant dusts have found some application as a protectant for stored grains and seeds (Highley *et al.*, 1994). These act by causing physical damage to the moisture barrier of the integument of a pest resulting in death through desiccation. Two groups of compounds are in common use: the activated silica dusts (e.g. Dryacide®) that are possibly used most widely; and the diatomaceous earths. A problem that can arise with the use of dusts on grain is that they affect the flow rate of grain streams and increase the power required for auguring.

Pesticides for Cut Flowers and Foliage

Cut flowers and foliage are a special category with regard to pesticide use because although they are fresh, actively respiring commodities with limited shelf life, the fact that they are not for consumption relaxes pesticide restrictions on them compared with fruits and vegetables.

Hansen and Hara (1994) review much of the literature from 1973 to 1993 concerning research on pesticidal dips and sprays for phytosanitary control on cut flowers and foliage and tolerance of the plants to the chemicals. In several cases insecticidal soaps showed as much promise as traditional organophosphorus pesticides.

Hata *et al.* (1993) tested the synthetic pyrethroids fluvalinate (with and without piperonyl butoxide) and cyfluthrin, the organophosphorus chlorpyrifos and abamectin as field sprays and postharvest dips against the thrips *Frankliniella occidentalis* and *Thrips palmi* on orchids. A double dip of chlorpyrifos or abamectin after harvest gave results likely to be acceptable at the levels required of a phytosanitary treatment for cut flowers where a disease vector role in the thrips is not a factor. Some reduction in vase life was recorded, but at levels probably acceptable in commercial trade.

Commercial Application

Application of pesticides on a commercial scale can be accomplished by spray or immersion or a dry formulation for stored products. Use of pesticides is generally one of the more economical and easy to apply phytosanitary treatments and well suited to commercial operations of all sizes. Some of the newer pesticides, however, can be quite expensive.

Fruit and vegetables

In commercial operations insecticide treatment is typically done by in-line spraying as part of the packing line operation (Fig. 12.1), but it can be done initially by immersing fruit held in a metal or plastic basket into a tank of insecticide of the required concentration. For quantities of about $1m^3$, a forklift is used; smaller quantities can be dipped manually if safe handling procedures are observed. The fruit or vegetable commodity to be treated should be washed, dried, culled and graded prior to dipping because the pesticide must be the last treatment applied. No chemical contaminant that could cause breakdown of the insecticide (e.g. sodium hypochlorite in an organophosphorus mix) should be allowed to enter the treatment solution. Accurate timing of immersion ensures that efficacy is achieved without exceeding the MRL. Both methods are most widely used in Queensland for tomatoes exported to fruit fly-free Australian states and New Zealand and were widely used on fruit from the quarantined zone as an emergency measure during eradication of an incursion by the Malaysian papaya fly, *Bactrocera papayae*, in northern Queensland. For the most commonly used insecticide, dimethoate, the maximum life of a dip solution should not exceed 1 month. During this time, concentrate is to be added every 7 days to compensate for decay of the active ingredient in accordance with its half-life of 148 days at pH 6 and 25°C (Noble, 1983).

Cut flowers and foliage

Dipping or in-line flood-spray application would be the usual methods of application. Hara *et al.* (1993) found that for thrips on orchids, double dipping a few hours apart was most effective owing to differing susceptibilities in life cycle stages with eggs being possibly the most tolerant. This is supported by the results of combination treatment experiments by Bansiddhi *et al.* (2004). For cut flowers, commercially prepared dichlorvos-impregnated resin strips are sometimes included in packages to enable disinfestation during transport.

Grains and seeds

Pesticides application to grains and seeds needs to be done in ways that cause little or no increase in moisture content. Where the purpose is as a combined grain or seed disinfestant and protectant, a dust formulation can be used but it can have the disadvantage of affecting grain flow behaviour in bulk handling necessitating increased power input to augers used in bulk storages. The most efficient mode of application is by way of a low volume water-based spray to a grain stream or auger at rates such as 1 l/t which has no significant effect on the moisture content of the commodity. Adequate dispersal occurs in the following grain stream movement.

Dichlorvos is more versatile than the other pesticides commonly used for grain and seed applications. It also has some fumigant action and can be applied repetitively in storage airspace as an aerosol, timed to coordinate with a susceptible pest stage such as the newly emerged adults of grain moths in storages, or continuously from slow-release impregnated resin strips (Bengston, 1976).

Combinations with Other Treatments

Combining insecticides with other treatments to disinfest fruits and vegetables has not been fully exploited. Hot water dips for disease control combined with low postharvest storage temperatures cause mortality to fruit flies (Armstrong and Couey, 1989). The addition of an insecticide to the hot water dip may provide quarantine security, although only thermostable insecticides can be used in this way. Insecticide added to hot water for disease control or applied separately, if necessary, could be used to increase mortality at postharvest storage temperatures to meet many overall disinfestation security requirements (Heather *et al.*, 1987). Bansiddhi *et al.* (2004) tested the insecticide imidachlorprid in combination with irradiation and storage at 15°C against *T. palmi* on orchids. None of these treatments alone was effective at levels tolerated by the flowers.

Other Applications

Dips can be used to disinfest planting stock, where because residues are not a problem for plant material that is not consumed, insecticides of higher risk can be used. In Australia, diazinon is used to disinfest citrus nursery stock of leafminer, *Phyllocnistis citrella*, and on various plant materials against cattle tick, *Boophilus microplus* (Anon., 1982). Field applications of 3000 l/ha of a 0.5 g/l spray of chlorpyrifos or diazinon have been used to disinfest pineapples of pineapple mealybug, *Dysmicoccus brevipes* (Beavis *et al.*, 1991), thus enabling the New Zealand quarantine maximum pest limit of 0.5% for that species to be met. Foliar applications, combined with cultural practices such as skirt pruning, provide quarantine security for citrus exported to Japan without sacrificing integrated pest management programmes which minimize pesticide residues on the fruit.

Conclusions

Where the criteria of safety, efficacy and practicality can be met, pesticides offer realistic treatments for postharvest quarantine disinfestation. The major disadvantages are residues and toxicity risks to operators. The advantages of insecticides include economical costs, logistical flexibility, simplicity of application, residual protection and ease of supervision. They are particularly appropriate where the efficacy required is less than 99.9968% and may achieve some accepted levels of quarantine security when applied preharvest. They can also be an important component of combination systems which, in total, meet efficacy requirements of 99.9968% or other levels.

There is a role for pesticides in both postharvest disinfestation and pest suppression in the field as adjuncts to phytosanitary systems, which include pest management systems to meet established MRL. Both usages are subject to the existence of achievable MRL. Where safety margins are adequate, some adjustment of MRL may be justified given their relationship to good agricultural practice rather than to public health risk alone, provided that witholding periods are observed to enable residues to decay to levels below the MRL. If approvals exist for field use of the pesticides used postharvest, no different residues are then being introduced to the product. Residues are more accurately controlled in postharvest application through precise concentration and timing of the dip or spray compared with field application, so postharvest usage can represent a safer usage of chemical pesticides. In Australia, the National Residue Survey (Anon., 1989) recorded a majority of market samples with no detectable residues, a situation expected to be true for most agricultural and horticultural production systems. However, with the exception of certified 'organic' produce, few edible plant products marketed today are entirely free of traces of pesticide residues, an accepted and safe trade-off for quality levels required by markets.

13 Miscellaneous Phytosanitary Treatments

An assortment of treatments both applied and only researched, do not fit into any of the major chapters covered thus far and do not comprise enough material by themselves to warrant entire chapters. They are discussed briefly in this chapter.

Cleaning and Pest Removal

Physical removal of surface pests from commodities has been a long-standing method of phytosanitary disinfestation. It is not a prescribed treatment that by itself provides a certain level of confidence that pest population levels are reduced to negligible levels, but efficacy must be verified by inspection. If live pests are found the load is usually rejected and must be fumigated or otherwise treated, returned or destroyed.

For cleaning to be effective the surface of a commodity should be tolerant to the rigours of the process and not provide many areas difficult to clean. For example, grapefruit would be much easier to clean than grapes. The most difficult surface pests to remove from fresh commodities are armoured scales (Diaspididae). Walker *et al.* (1996) removed up to 98% of red scale, *Aonidiella aurantii*, from navel orange using a commercial system of high-pressure water nozzles and brushes on a packing line. Pressure at 3.6 MPa for about 20 s caused some rind injury and discoloration while 2.9 MPa did not. Whiting *et al.* (1998a) found that high pressure was not very effective at removing three armoured scale species from kiwi fruit. The best treatment (30 s in 55°C water followed by 13.8 MPa for 1 s and 3 min of mild brushing) cleaned 17% of the fruit and killed 94% of the armoured scales. The 55°C hot water immersion by itself killed 88% of the scales while a 30 s immersion at 65°C killed 100%, but did not remove them. A treatment for kiwi fruit to Japan would need to remove armoured scales as it is too difficult to determine if they are alive.

High-pressure (5.5 and 3.4 MPa) washing on a packing line removed a large amount of lightbrown apple moth, *Epiphyas postvittana*, eggs and early instars from apples (Whiting *et al.*, 1998b). However, once larvae began tunnelling into the fruit the washing was not very effective. With the same system complete removal of mealybugs, *Pseudococcus viburni*, did not occur, and only 12% of mealybugs under the calyx were removed by the highest pressure.

Hansen *et al.* (2006) found that increasing water pressure beyond 420 kPa did not increase removal of a mealybug, a spider mite and an aphid from apples and pears. Complete removal was not obtained. The high-pressure washing system at 400 kPa for 15 s removed 90% of codling moth, *Cydia pomonella*, and 60% of European red mite eggs from apples and pears (Neven *et al.*, 2006a).

Coating as a Physical Treatment

Edible coatings that restrict gaseous exchange modify the atmosphere inside coated commodities, leading to mortality of fruit flies (Chapter 11). The effect of coatings on small surface pests is physical, binding the pest and gluing it in place. Washing followed by coating is an approved treatment against the mite *Brevipalpus chilensis* on cherimoya, lime and passion fruit from Chile to the USA (APHIS, 2007). Because complete coating of the fruit is essential the import inspector is advised to check for it.

This treatment should be considered for other phytosanitary problems with small arthropod pests especially where coatings are already used on the quarantined commodities to preserve quality.

Pressure

Baling hay to a pressure of 10.3 MPa for 24 h killed 100% of cereal leaf beetle, *Oulema melanopus* (Yokoyama and Miller, 2002). However, Canada requires that the treatment be done along with phosphine fumigation (2.1 g/m^3 for 3 days at $\geq 21°C$) for export to uninfested regions of that country. Japan accepts a combination pressure/phosphine treatment for control of Hessian fly, *Mayetiola destructor*, on hay shipped there (Yokoyama and Miller, 2003). Pressure alone may not suffice for a commodity like hay because small pockets could form in the pressurized matrix where insects are not killed.

Ultrasound

Hansen (2001) achieved about 80% control of a thrips and a spider mite on apples after 10 min exposure in an ultrasound machine. Another species of thrips was not appreciably controlled via two other ultrasound machines on asparagus spears after 8 min (van Epenhuijsen *et al.*, 1997). Ultrasound deserves some additional research with a broader investigation of parameters affecting efficacy, such as the amplitude of the sound waves. Because ultrasound is used to lyse cells to remove contents it might be expected to injure fresh commodities at doses required to kill quarantine organisms.

Pulsed Electric Field

Hallman and Zhang (1997) found that pulsed electric field in vitro inhibited the development of Mexican fruit fly, *Anastrepha ludens*, eggs and larvae. As few as three 50 μs pulses (2 kV/cm) prevented normal pupariation and adult emergence (Fig. 13.1). Eggs were more tolerant; ten 50 μs pulses at 5 kV/cm were required to prevent development to third instar. This technique requires considerable research before it could be considered as a commercial-scale phytosanitary treatment. Efficacy against pests on host material or effect on the host itself has not yet been studied.

A similar electromagnetic treatment killed 100% of New Zealand flower thrips, *Thrips obscuratus*, in some situations (van Epenhuijsen *et al.*, 2001). Some heating of the water in which the 2 min experiments were conducted occurred, but the final temperature (maximum 30°C) was below lethal temperatures, indicating that mortality was not caused by heat. The researchers suggested that the mode of action could be a combination of electric field intensity, electrohydraulic shock and ozone poisoning.

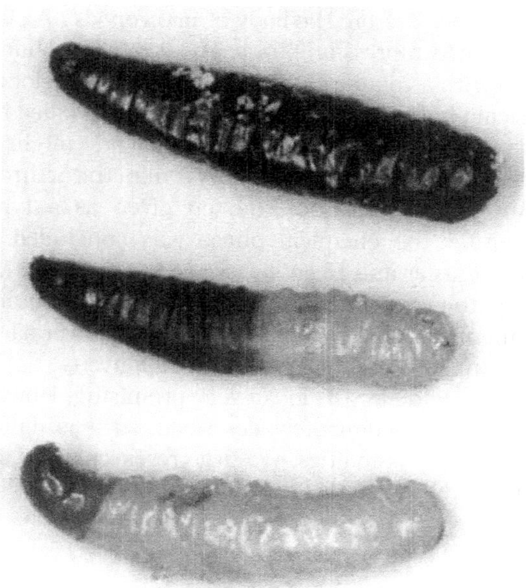

Fig. 13.1. Various degrees of larviform pupariation of Mexican fruit fly third instars subjected to pulsed electric field (Source: Hallman and Zhang, 1997).

Atmospheric Pressure Plasma Discharge

Donohue *et al.* (2006) and Bures *et al.* (2005, 2006) tested an atmospheric pressure plasma discharge on several insects in vitro. Green peach aphid, *Myzus persicae*, Asian tiger mosquito, *Aedes albopictus*, and human body louse, *Pediculus humanus humanus*, were highly susceptible; the German cockroach, *Blattella germanica*, was considerably tolerant; and two species of thrips and one spider mite were intermediate. The mode of action of plasma on insects was found to be an interaction with the nervous or neuromuscular system. Preliminary studies with tobacco leaves showed injury to the leaves 24 h after treatment at 60 s as well as a reduction in the effectiveness of the discharge on green peach aphid mortality when on tobacco leaves. Considerable research is needed before this technique could be evaluated as a commercial phytosanitary treatment.

Metabolic Stress Disinfection and Disinfestation

Metabolic stress disinfection and disinfestation (MSDD) is defined as 'a combination of physical and chemical techniques (phases) that use a rapid sequence of mechanical forces' to 'create an extremely low oxygen environment' (Lagunas-Solar *et al.*, 2006). 'This effect disrupts respiration ... while causing irreversible shifts in cellular chemistry.' The MSDD process starts with a series (about ten) of rapid (< 1 min) decompression and compression cycles. The compression cycles use carbon dioxide. At the end of the cycles 'oxygen is essentially eliminated ... from the body of arthropods'. A chemical phase follows the decompression/compression phase. The treatment chamber is decompressed again and filled with a volatile chemical such as ethanol for about 3–4 h.

The physical phase alone carried on for 6–12 h killed between 80–100% of several arthropods. It is not clearly stated, but one might assume that if each decompression/compression cycle lasted < 1 min, the arthropods must have gone through > 360–720 cycles, not the ten given as a typical number in the methodology. After the chemical phase was concluded, all stages tested of *Drosophila melanogaster* and *Heliothis virescens* were dead. MSDD also controlled several plant pathogenic microorganisms, although detailed methodology on how the treatment was applied is lacking. Injury to several fresh commodities was light, but the actual process they received is not given.

Taken at face value MSDD looks very promising. However, the experiments published so far are incompletely described, so it is difficult to come to any conclusion about its possibilities. We urge researchers to continue studies with this process and publish detailed methods and results.

Other Techniques

There are other methods that might be studied for their use as phytosanitary treatments. Many methods being studied in food preservation, such as oscillating magnetic fields, could have some benefit in phytosanitary treatments. Some novel treatments appear in brief summaries at the Annual International Research Conference on Methyl Bromide Alternatives and Emissions Reductions (MBAO, 2007). Of particular interest might be techniques that attack systems that pests possess but not their hosts, such as nervous systems. Because arthropods are more complex than the microorganisms of concern in food technology and the plant host commodities they infest, it may be easier to control them. For example, inactivation of microorganisms commonly requires pulsed electric fields of 25–60 kV/cm (Barsotti and Cheftel, 1999), while the Mexican fruit fly was controlled with 5 kV/cm (Hallman and Zhang, 1997).

Appendix I: the IPPC Model Phytosanitary Certificate

This is a model phytosanitary certificate (from ISPM 12; IPPC, 2007). Phytosanitary certificates are required widely in trade and have been in use since early in the 19th century with varying degrees of acceptance. Levels of detail differ from country to country and according to specific requirements. Recent efforts by IPPC have increased their acceptance through both higher reliability and harmonization of format.

Model Phytosanitary Certificate

No. ________

Plant Protection Organization of ______________________________
TO: Plant Protection Organization(s) of ______________________________

I. Description of Consignment

Name and address of exporter: ______________________________
Declared name and address of consignee: ______________________________
Number and description of packages: ______________________________
Distinguishing marks: ______________________________
Place of origin: ______________________________
Declared means of conveyance: ______________________________
Declared point of entry: ______________________________
Name of produce and quantity declared: ______________________________
Botanical name of plants: ______________________________

This is to certify that the plants, plant products or other regulated articles described herein have been inspected and/or tested according to appropriate official procedures and are considered to be free from the quarantine pests specified by the importing contracting party and to conform with the current phytosanitary requirements of the importing contracting party, including those for regulated non-quarantine pests.

They are deemed to be practically free from other pests.*

II. Additional Declaration

III. Disinfestation and/or Disinfection Treatment

Date ________ Treatment __________ Chemical (active ingredient) ______________
Duration and temperature ______________________________
Concentration ______________________________
Additional information ______________________________
____

Place of issue ______________________________

(Stamp of Organization) Name of authorized officer ______________

Date ______________ (Signature) ______________________

No financial liability with respect to this certificate shall attach to (name of Plant Protection Organization) or to any of its officers or representatives.*

* Optional clause

Appendix II: EPPO Guidelines on Pest Risk Analysis (EPPO, 2006)

European and Mediterranean Plant Protection Organization
Organisation Européenne et Méditerranéenne pour la Protection des Plantes

Guidelines on Pest Risk Analysis
Lignes directrices pour l'analyse du risque phytosanitaire

Pest Risk Analysis record format for PM5/3 (2) *Decision-support scheme for quarantine pests* (version 2006-09)

PEST RISK ANALYSIS FOR			
Pest risk analyst(s):			
			WARNING: **This template should be used together with PM5/3 (2) *Decision-support scheme for quarantine pests* which contains the necessary explanations for a proper understanding of the scheme** **Some questions of PM5/3 (2) are subdivided in the template (e.g. 2, 2a and 2b) for practical reasons** **Similarly, some instructions present in the introduction text of different sections of the decision support scheme are presented as questions in the template (e.g. 1.19A corresponding to the introduction of the section '*suitability of environment*').**
Date:			
Stage 1: Initiation			
1 What is the reason for performing the PRA?			
2 Enter the name of the pest			
2a Indicate the type of the pest			
2b Indicate the taxonomic position			
3 Clearly define the PRA area			
4 Does a relevant earlier PRA exist?			
5 Is the earlier PRA still entirely valid, or only partly valid (out of date, applied in different circumstances, for a similar but distinct pest, for another area with similar conditions)?			

Stage 2A: Pest Risk Assessment – Pest categorization		
<u>Identify the pest (or potential pest)</u>		
6 Does the name you have given for the organism correspond to a single taxonomic entity which can be adequately distinguished from other entities of the same rank?		
7 Even if the causal agent of particular symptoms has not yet been fully identified, has it been shown to produce consistent symptoms and to be transmissible?		
<u>Determining whether the organism is a pest</u>		
8 Is the organism in its area of current distribution a known pest (or vector of a pest) of plants or plant products?		
9 Does the organism have intrinsic attributes that indicate that it could cause significant harm to plants?		
<u>Presence or absence in the PRA area and regulatory status (pest status)</u>		
10 Does the pest occur in the PRA area?		
11 Is the pest widely distributed in the PRA area?		
<u>Potential for establishment and spread in the PRA area</u>		
12 Does at least one host-plant species (for pests directly affecting plants) or one suitable habitat (for non-parasitic plants) occur in the PRA area (outdoors, in protected cultivation or both)?		
13 If a vector is the only means by which the pest can spread, is a vector present in the PRA area? (If a vector is not needed or is not the only means by which the pest can spread go to 14.)		

14 Does the known area of current distribution of the pest include ecoclimatic conditions comparable with those of the PRA area or sufficiently similar for the pest to survive and thrive (consider also protected conditions)?		
<u>Potential for economic consequences in PRA area</u>		
15 With specific reference to the plant(s) or habitats which occur(s) in the PRA area, and the damage or loss caused by the pest in its area of current distribution, could the pest by itself, or acting as a vector, cause significant damage or loss to plants or other negative economic impacts (on the environment, on society, on export markets)?		
<u>Conclusion of pest categorization</u>		
16 This pest could present a risk to the PRA area.		
17 The pest does not qualify as a quarantine pest for the PRA area and the assessment for this pest can stop (summarize the main reason for stopping the analysis).		
Section 2B: Pest Risk Assessment – Probability of introduction/spread and of potential economic consequences		
1. Probability of introduction Introduction, as defined by the FAO Glossary of Phytosanitary Terms, is the entry of a pest resulting in its establishment.		
Probability of entry of a pest		
<u>Identification of pathways</u> Note: If the most important pathway is intentional import, do not consider entry,		

but go directly to establishment. Spread from the intended habitat to the unintended habitat, which is an important judgement for intentionally imported organisms, is covered by questions 1.33 and 1.35.		
1.1 Consider all relevant pathways and list them.		
1.2 Estimate the number of relevant pathways, of different commodities, from different origins, to different end uses.		
1.3 Select from the relevant pathways, using expert judgement, those which appear most important. If these pathways involve different origins and end uses, it is sufficient to consider only the realistic worst-case pathways. The following group of questions on pathways is then considered for each relevant pathway in turn, as appropriate, starting with the most important.		
Pathway n°:		Repeat this section for all relevant pathways
Probability of the pest being associated with the individual pathway at origin		
1.4 How likely is the pest to be associated with the pathway at origin?		
1.5 Is the concentration of the pest on the pathway at origin likely to be high, taking into account factors like cultivation practices, treatment of consignments?		
1.6 How large is the volume of the movement along the pathway?		
1.7 How frequent is the movement along the pathway?		

Probability of survival during transport or storage		
1.8 How likely is the pest to survive during transport/storage?		
1.9 How likely is the pest to multiply/increase in prevalence during transport/storage?		
Probability of the pest surviving existing pest management procedures		
1.10 How likely is the pest to survive or remain undetected during existing phytosanitary measures?		
1.11 In the case of a commodity pathway, how widely is the commodity to be distributed throughout the PRA area?		
1.12 In the case of a commodity pathway, do consignments arrive at a suitable time of year for pest establishment?		
1.13 How likely is the pest to be able to transfer from the pathway to a suitable host or habitat?		
1.14 In the case of a commodity pathway, how likely is the intended use of the commodity (e.g. processing, consumption, planting, disposal of waste, by-products) to aid transfer to a suitable host or habitat?		
Consideration of further pathways		
1.15 Do other pathways need to be considered?		
Conclusion on the probability of entry		
The overall probability of entry should be described and risks presented by different pathways should be identified.		

Probability of establishment		
Availability of suitable hosts or suitable habitats, alternate hosts and vectors in the PRA area		
1.16a Specify the host plant species (for pests directly affecting plants) or suitable habitats (for non-parasitic plants) present in the PRA area.		
1.16b Estimate the number of host plant species or suitable habitats in the PRA area.		
1.17 How widespread are the host plants or suitable habitats in the PRA area? (specify)		
1.18 If an alternate host is needed to complete the life cycle, how widespread are alternate host plants in the PRA area?		
1.19 If the pest requires another species for critical stages in its life cycle such as transmission (e.g. vectors), growth (e.g. root symbionts), reproduction (e.g. pollinators) or spread (e.g. seed dispersers), how likely is the pest to become associated with such species?		
Suitability of the environment		
1.19A Specify the area where host plants (for pests directly affecting plants) or suitable habitats (for non-parasitic plants) are present (cf. QQ 1.16–1.19). This is the area for which the environment is to be assessed in this section. If this area is much smaller than the PRA area, this fact will be used in defining the endangered area.		
1.20 How similar are the climatic conditions that would affect pest establishment, in the PRA area and in the current area of distribution?		

1.21 How similar are other abiotic factors that would affect pest establishment, in the PRA area and in the current area of distribution?		
1.22 If protected cultivation is important in the PRA area, how often has the pest been recorded on crops in protected cultivation elsewhere?		
1.23 How likely is it that establishment will not be prevented by competition from existing species in the PRA area?		
1.24 How likely is it that establishment will not be prevented by natural enemies already present in the PRA area?		
<u>Cultural practices and control measures</u>		
1.25 To what extent is the managed environment in the PRA area favourable for establishment?		
1.26 How likely is it that existing control or husbandry measures will fail to prevent establishment of the pest?		
1.27 How likely is it that the pest could survive eradication programmes in the PRA area?		
<u>Other characteristics of the pest affecting the probability of establishment</u>		
1.28 How likely is the reproductive strategy of the pest and the duration of its life cycle to aid establishment?		
1.29 How likely are relatively small populations or populations of low genetic diversity to become established?		

1.30 How adaptable is the pest? Adaptability is:		
1.31 How often has the pest been introduced into new areas outside its original area of distribution? (specify the instances, if possible)		
1.32 Even if permanent establishment of the pest is unlikely, how likely are transient populations to occur in the PRA area through natural migration or entry through man's activities (including intentional release into the environment)?		
<u>Conclusion on the probability of establishment</u>		
The overall probability of establishment should be described.		
Probability of spread		
1.33 How likely is the pest to spread rapidly in the PRA area by natural means?		
1.34 How likely is the pest to spread rapidly in the PRA area by human assistance?		
1.35 How likely is it that the spread of the pest will not be contained within the PRA area?		
<u>Conclusion on the probability of spread</u>		
The overall probability of spread should be described.		
Conclusion on the probability of introduction and spread		
The overall probability of introduction and spread should be described. The probability of introduction and spread may be expressed by comparison with PRAs on other pests.		

Conclusion regarding endangered areas		
1.36 Based on the answers to questions 1.16 to 1.35 identify the part of the PRA area where presence of host plants or suitable habitats and ecological factors favour the establishment and spread of the pest to define the endangered area.		
2. Assessment of potential economic consequences		
2.0 In any case, providing replies for all hosts (or all habitats) and all situations may be laborious, and it is desirable to focus the assessment as much as possible. The study of a single worst-case may be sufficient. Alternatively, it may be appropriate to consider all hosts/habitats together in answering the questions once. Only in certain circumstances will it be necessary to answer the questions separately for specific hosts/habitats.		
Pest effects		
2.1 How great a negative effect does the pest have on crop yield and/or quality to cultivated plants or on control costs within its current area of distribution?		
2.2 How great a negative effect is the pest likely to have on crop yield and/or quality in the PRA area?		
2.3 How great an increase in production costs (including control costs) is likely to be caused by the pest in the PRA area?		
2.4 How great a reduction in consumer demand is the pest likely to cause in the PRA area?		

2.5 How important is environmental damage caused by the pest within its current area of distribution?		
2.6 How important is the environmental damage likely to be in the PRA area?		
2.7 How important is social damage caused by the pest within its current area of distribution?		
2.8 How important is the social damage likely to be in the PRA area?		
2.9 How likely is the presence of the pest in the PRA area to cause losses in export markets?		
2.9A As noted in the introduction to section 2, the evaluation of the following questions may not be necessary if any of the responses to questions 2.2, 2.3, 2.4, 2.6, 2.8 or 2.9 is 'major or massive' or 'very likely' or 'certain'. You may go directly to point 2.16 unless a detailed study of impacts is required.		
2.10 How easily can the pest be controlled in the PRA area?		
2.11 How likely is it that natural enemies, already present in the PRA area, will not suppress populations of the pest if introduced?		
2.12 How likely are control measures to disrupt existing biological or integrated systems for control of other pests or to have negative effects on the environment?		
2.13 How important would other costs resulting from introduction be?		

2.14 How likely is it that genetic traits can be carried to other species, modifying their genetic nature and making them more serious plant pests?		
2.15 How likely is the pest to act as a vector or host for other pests?		
2.15A Do you wish to consider the questions 2.1 to 2.15 again for further hosts/habitats?		
Conclusion of the assessment of economic consequences		
2.16 Referring back to the conclusion on endangered area (1.36), identify the parts of the PRA area where the pest can establish and which are economically most at risk.		
Degree of uncertainty		
Estimation of the probability of introduction of a pest and of its economic consequences involves many uncertainties. In particular, this estimation is an extrapolation from the situation where the pest occurs to the hypothetical situation in the PRA area. It is important to document the areas of uncertainty and the degree of uncertainty in the assessment, and to indicate where expert judgement has been used. This is necessary for transparency and may also be useful for identifying and prioritizing research needs. It should be noted that the assessment of the probability and consequences of environmental hazards of pests of uncultivated plants often involves greater uncertainty than for pests of cultivated plants. This is due to the lack of information, additional complexity associated with ecosystems, and variability associated with pests, hosts or habitats.		

Conclusion of the pest risk assessment		
Entry Evaluate the probability of entry and indicate the elements which make entry most likely or those that make it least likely. Identify the pathways in order of risk and compare their importance in practice.		
Establishment Evaluate the probability of establishment, and indicate the elements which make establishment most likely or those that make it least likely. Specify which part of the PRA area presents the greatest risk of establishment.		
Economic importance List the most important potential economic impacts, and estimate how likely they are to arise in the PRA area. Specify which part of the PRA area is economically most at risk.		
Overall conclusion of the pest risk assessment The risk assessor should give an overall conclusion on the pest risk assessment and an opinion as to whether the pest or pathway assessed is an appropriate candidate for stage 3 of the PRA: the selection of risk management options, and an estimation of the pest risk associated.		
This is the end of the pest risk assessment		

Stage 3: Pest Risk Management		
3.1 Is the risk identified in the Pest Risk Assessment stage for all pest/pathway combination an acceptable risk?		
Pathway n°:		Repeat this section for all relevant pathways
3.2 Is the pathway that is being considered a commodity of plants and plant products?		
3.3 Is the pathway that is being considered the natural spread of the pest? (see answer to question 1.33)		
3.4 Is the pest already entering the PRA area by natural spread or likely to enter in the immediate future? (see answer to question 1.33)		
3.5 Could entry by natural spread be reduced or eliminated by control measures applied in the area of origin?		
3.6 Could the pest be effectively contained or eradicated after entry? (see answer to questions 1.27, 1.35)		
3.7 Was the answer 'yes' to either question 3.5 or question 3.6?		
3.8 Is the pathway that is being considered the entry with human travellers?		
3.9 Is the pathway being considered contaminated machinery or means of transport?		
Existing phytosanitary measures		
3.10 Are there any existing phytosanitary measures applied on the pathway that could prevent the introduction of the pest?		

Identification of appropriate risk management options		
Options for consignments		
Detection of the pest in consignments by inspection or testing		
3.11 Can the pest be reliably detected by a visual inspection of a consignment at the time of export during transport/storage or at import?		
3.12 Can the pest be reliably detected by testing (e.g. for pest plant, seeds in a consignment)?		
3.13 Can the pest be reliably detected during post-entry quarantine?		
Removal of the pest from the consignment by treatment or other phytosanitary procedures		
3.14 Can the pest be effectively destroyed in the consignment by treatment (chemical, thermal, irradiation, physical)?		
3.15 Does the pest occur only on certain parts of the plant or plant products (e.g. bark, flowers), which can be removed without reducing the value of the consignment? (This question is not relevant for pest plants.)		
3.16 Can infestation of the consignment be reliably prevented by handling and packing methods?		
Prevention of establishment by limiting the use of the consignment		
3.17 Could consignments that may be infested be accepted without risk for certain end uses, limited distribution in the PRA area, or limited periods of entry, and can such limitations be applied in practice?		

Options for the prevention or reduction of infestation in the crop		
Prevention of infestation of the commodity		
3.18 Can infestation of the commodity be reliably prevented by treatment of the crop?		
3.19 Can infestation of the commodity be reliably prevented by growing resistant cultivars? (This question is not relevant for pest plants.)		
3.20 Can infestation of the commodity be reliably prevented by growing the crop in specified conditions (e.g. protected conditions such as screened greenhouses, physical isolation, sterilized growing medium, exclusion of running water . . .)?		
3.21 Can infestation of the commodity be reliably prevented by harvesting only at certain times of the year, at specific crop ages or growth stages?		
3.22 Can infestation of the commodity be reliably prevented by production in a certification scheme (i.e. official scheme for the production of healthy plants for planting)?		
Establishment and maintenance of pest freedom of a crop, place of production or area		
3.23 Has the pest a very low capacity for natural spread?		
3.24 Has the pest a low to medium capacity for natural spread?		
3.25 Has the pest a medium capacity for natural spread?		
3.26 The pest is of medium to high capacity for natural spread.		

3.27 Can pest freedom of the crop, place of production or an area be reliably guaranteed?		
Consideration of other possible measures		
3.28 Are there effective measures that could be taken in the importing country (surveillance, eradication) to prevent establishment and/or economic or other impacts?		
Evaluation of risk management options		
3.29 Have any measures been identified during the present analysis that will reduce the risk of introduction of the pest?		
3.30 Taking each of the measures identified individually, does any measure on its own reduce the risk to an acceptable level?		
3.31 For those measures that do not reduce the risk to an acceptable level, can two or more measures be combined to reduce the risk to an acceptable level?		
3.32 If the only measures available reduce the risk but not down to an acceptable level, such measures may still be applied, as they may at least delay the introduction or spread of the pest. In this case, a combination of phytosanitary measures at or before export and internal measures (see question 3.29) should be considered.		
3.33 Estimate to what extent the measures (or combination of measures) being considered interfere with international trade.		

3.34 Estimate to what extent the measures (or combination of measures) being considered are cost-effective, or have undesirable social or environmental consequences.		
3.35 Have measures (or combination of measures) been identified that reduce the risk for this pathway, and do not unduly interfere with international trade, are cost-effective and have no undesirable social or environmental consequences?		
3.36 Envisage prohibiting the pathway.		
3.37 Have all major pathways been analysed (for a pest-initiated analysis)?		
3.38 Have all the pests been analysed (for a pathway-initiated analysis)?		
3.39 For a pathway-initiated analysis, compare the measures appropriate for all the pests identified for the pathway that would qualify as quarantine pests, and select only those that provide phytosanitary security against all the pests.		
3.40 Consider the relative importance of the pathways identified in the conclusion to the entry section of the pest risk assessment.		
3.41. All the measures identified as being appropriate for each pathway or for the commodity can be considered for inclusion in phytosanitary regulations in order to offer a choice of different measures to trading partners.		

3.42 In addition to the measure(s) selected to be applied by the exporting country, a phytosanitary certificate (PC) may be required for certain commodities. The PC is an attestation by the exporting country that the requirements of the importing country have been fulfilled. In certain circumstances, an additional declaration on the PC may be needed (see EPPO Standard PM 1/1(2): Use of phytosanitary certificates).		
3.43 If there are no measures that reduce the risk for a pathway, or if the only effective measures unduly interfere with international trade (e.g. prohibition), are not cost-effective or have undesirable social or environmental consequences, the conclusion of the pest risk management stage may be that introduction cannot be prevented.		
Conclusion of Pest Risk Management Summarize the conclusions of the Pest Risk Management stage. List all potential management options and indicate their effectiveness. Uncertainties should be identified.		

Glossary

ADI: acceptable daily intake (e.g. of food additive or pesticide residue)

APHIS: Animal and Plant Health Inspection Service of US Department of Agriculture (USDA)

APPPC: Asia and Pacific Plant Protection Commission

AQIS: Australian Quarantine and Inspection Service area freedom: having pest free area (PFA) status

ARS: Agricultural Research Service of the US Department of Agriculture

C: Celsius or centigrade, the SI scale of temperature measurement

CA: Comunidad Andina

CA: controlled storage atmosphere for foods with ongoing management of the atmosphere constituents

CATTS: controlled atmosphere temperature treatment system for disinfestation of fresh fruits and vegetables

CBD: Convention on Biological Diversity

CL: Confidence level or limits for statistical data, customarily 0.95; *see also* entry for FL (fiducial limits)

CLIMEX: a powerful geographic information system (GIS) model centred on climate that is frequently used for comparative studies

COSAVE: Comité Regional de Sanidad Vegetal para el Cono Sur

CPM: Commission on Phytosanitary Measures

CPPC: Caribbean Plant Protection Commission

CSIRO: Commonwealth Scientific and Industrial Research Organization, Australia

CT: concentration × time product which defines a fumigant dose

DUR: Dose uniformity ratio of gamma- or X-radiation also known as max:min ratio

EIL: economic injury level, the level of pest injury at which economic loss occurs and the cost of pest management is justified

EMC: equilibrium moisture content, related to relative humidity (RH)

EPA: Environmental Protection Agency (or Authority) of the USA and other governments
EPPO: European and Mediterranean Plant Protection Organization
EPR: electron paramagnetic resonance
F: Fahrenheit, the imperial scale of temperature measurement
FAO: Food and Agriculture Organization of the United Nations
FDA: Food and Drug Administration, US government
FL: fiducial limits; these relate to statistical data, typically defining the confidence belt containing the mean value or regression line
GATT: General Agreement on Tariffs and Trade
GIS: geographic information system, a computer system which collates, stores and integrates complex environmental data according to location
GM: genetically modified as in transgenic crops or organisms
GMO: genetically modified organism
Gy: the SI unit of irradiation dose
IAEA: international Atomic Energy Agency
IAPSC: Inter-African Phytosanitary Council
ICGFI: International Consultative Group on Food Irradiation
IDIDAS: International Database on Insect Disinfestation and Sterilization sponsored jointly by the International Atomic Energy Agency (IAEA) and the Food and Agriculture Organization (FAO)
IPM: integrated pest management, a strategy to maximize cultural and biological inputs to pest control and to minimize any adverse effects of pesticide use
IPPC: International Plant Protection Convention, an international treaty hosted under the Food and Agriculture Organization (FAO)
ISO: International Standards Organization
ISPM: International Standard for Phytosanitary Measures of the International Plant Protection Convention (IPPC)
LC: lethal concentration, of a treatment; usually related to percentage mortality, e.g. LC_{50} is the lethal concentration of a treatment to achieve 50% mortality
LD: lethal dose, of a treatment usually related to percentage mortality, e.g. LD_{50} is the lethal dose of a treatment to achieve 50% mortality
LT: lethal time, of a treatment usually related to percentage mortality, e.g. LT_{50} is the lethal time of a treatment to achieve 50% mortality
MA: modified storage atmosphere usually for food commodities
MAF: Ministry of Agriculture and Forestry, New Zealand
MAFF: Ministry of Agriculture Forestry and Fisheries, Japan
max:min ratio: *see* DUR
MBAO: Methyl Bromide Alternatives Outreach
MC: moisture content; can be related to relative humidity (RH) as equilibrium moisture content (EMC)
MPL: maximum pest limit
MRL: maximum residue limits, set nationally or internationally by the Codex Alimentarius Committee of the World Health Organization (WHO)
MSDD: metabolic stress disinfection and disinfestation
NAPPO: North American Plant Protection Organization
NEPPO: Near East Plant Protection Organization

NHMRC: National Health and Medical Research Council, Australia
NPPO: national plant protection organization
NTP: normal temperature and pressure as it relates to the physical state of a chemical
OIRSA: Organismo Internacional Regional de Sanidad Agropecuario
PFA: pest free area
PMMA: polymethylmethacrylate or perspex as used in irradiation dosimetry
ppm: parts per million
PPPO: Pacific Plant Protection Organization
PRA: pest risk analysis, a structured appreciation of pest risk usually with relevance to import of agricultural produce
PRM: pest risk management, a part of pest risk analysis, usually with respect to importation of produce with phytosanitary risks
RF: radiofrequency; relevant to energy sources for heating disinfestation
RH: relative humidity
RPPO: regional plant protection organization
RSPM: Regional Standard for Phytosanitary Measures prepared by a Regional Plant Protection Organization (RPPO)
SE: standard error of the statistical mean
SI: Système International, the international system of metric measurement units
SIRM: sterile insect release method, a strategy used in pest eradication or suppression
SIT: sterile insect technique
SOM: self-organizing map, a system to determine species assemblages
SPS: Sanitary and Phytosanitary Agreement, a World Trade Organization (WTO) agreement
TDA: 9Z,tetradecadienyl acetate, a multi-species stored-product moth attractant
TLV: threshold limit value of a fumigant, an operator safety value
USDA: US Department of Agriculture
USGS: US Geological Survey
WHO: World Health Organization, a constituent organization of the United Nations Organization
WTO: World Trade Organization

References

Abbott, W.S. (1925) A method of computing the effectiveness of an insecticide. *Journal of Economic Entomology* 18, 265–267.

Adamo, M., D'llio, V., Gionfriddo, F., Nobili, P., Pasquali, A., Postorino, E., Rossi, G. and Zarbo, F. (1996) Le technologie di ionizzazione per frutti di arancio infestati da *Ceratitis capitata. L'Informatore Agrario* 52, 73–75.

Addo-Bediako, A., Tameru, B., Jackal, N. and Bonsi, C.K. (2007) Assessment of risk of introduction of *Cylas formicarius elegantulus* (Coleoptera: Brentidae) into weevil-free areas in the southern USA. *Journal of Economic Entomology* 100, 315–321.

AG (Australia Group) (2005) AG Common Control Lists: List of Plant Pathogens for Export Control. The Australia Group. Available at: http://www.australiagroup.net/en/control_list/plants.htm (accessed 25 June 2007).

Agnello, A.M., Spangler, S.M., Minson, E.S., Harris, T. and Kain, D.P. (2002) Effect of high-carbon dioxide atmospheres on infestations of apple maggot (Diptera: Tephritidae) in apples. *Journal of Economic Entomology* 95, 520–526.

Aharoni, Y., Stewart, J.K. and Guadagni, D.G. (1981) Modified atmospheres to control western flower thrips on harvested strawberries. *Journal of Economic Entomology* 74, 338–340.

Alibek, K. (1999) The Soviet Union's anti-agricultural biological weapons. *Annals of the New York Academy of Sciences* 894, 18–19.

Allwood, A.J., Chinajariyawong, A., Drew, R.A.I., Hamacek, E.J., Hancock, D.L., Hengsawad, C., Jipanin, J.C., Jirasurat, M., Kong Krong, C., Kristanneepaiboom, S., Leong, C.T.S. and Vijaysegaran, S. (1999) Host plant records for fruit flies (Diptera: Tephritidae) in South East Asia. *The Raffles Bulletin of Zoology* Supplement 7, 92 pp.

Alonso, M., del Río, M.A. and Jacas, J. (2002) Respuesta del híbrido de mandarina 'Ellendale' infestado con *Ceratitis capitata* (Wiedemann) (Diptera: Tephritidae) al tratamiento con atmósfera insecticida y calor. *Boletín de Sanidad Vegetal de Plagas* 28, 427–433.

Alonso, M., Del Río, M.A. and Jacas, J.-A. (2005) Carbon dioxide diminishes cold tolerance of third instar larvae of *Ceratitis capitata* Wiedemann (Diptera: Tephritidae) in 'Fortune' mandarins: implications for citrus quarantine treatments. *Postharvest Biology and Technology* 36, 103–111.

Amerine, M.A., Pangborn, R.M. and Roessler, E.B. (1965) *Principles of Sensory Evaluation.* Academic Press, New York.

Anderson, T.E. and Leppla, N.C. (eds) (1992) *Advances in Insect Rearing for Research and Pest Management.* Westview Press, Boulder, Colorado.

Anon. (1982) *Statutory Rules, Vegetation and Vine Diseases Act (State of Victoria).* Government Printer, Melbourne, Victoria, Australia.

Anon. (1989) *Guide to Import Plant Quarantine in Japan.* Ministry of Agriculture Forestry and Fisheries, Japan.

Anon. (1993) *Systems Operational Manual: Australia–New Zealand Bilateral Quarantine Agreement.* Australian Quarantine Inspection Service, Queensland, Australia.

Anon. (1994) Methyl bromide under fire. *Environmental Health Perspectives* 102, 732.

Anon. (2001a) *Specification for Determination of Fruit Fly Host Status as a Treatment. Regulatory Authority Standard 155.02.02.* Ministry of Agriculture and Forestry, Wellington, New Zealand.

Anon. (2001b) *Technical Review of the TriState Strategy for Queensland Fruit Fly.* Standing Committee on Agriculture and Resource Management, Department of Agriculture Fisheries and Forestry, Canberra, Australian Capital Territory, Australia.

Anon. (2002) *Predicting Invasions of Non-indigenous Plants and Plant Pests.* Committee on the Scientific Basis for Predicting the Invasive Potential of Nonindigenous Plants and Plant Pests in the USA. National Academy Press, Washington, DC.

Anon. (2005) Requirements for establishment of pest free area for mango nut (seed) weevil (*Sternochetus mangiferae*) and pulp weevil (*S. frigidus*). *National Standard(s) for Phytosanitary Measures (NSPM) – 13.* Ministry of Agriculture, Government of India, Faridabad, India.

Anon. (2006a) *Understanding and Responding to Climate Change.* The National Academies, Washington, DC.

Anon. (2006b) Papaya firms struggling for answers. *The Packer,* 26 June, 113, A1–2.

Anon. (2007a) GENSTAT. VSN International, UK. Available at: http://www.vsni.co.uk/products/genstat/ (accessed 25 June 2007).

Anon. (2007b) SAS. SAS Institute, USA. Available at: http://www.sas.com/ (accessed 25 June 2007).

APHIS (US Department of Agriculture Animal and Plant Health Inspection Service) (2000a) Guidelines for Pathway-initiated Pest Risk Assessments. Available at: http://www.aphis.usda.gov/ppq/pra/ (accessed 3 June 2005; not found 25 June 2007).

APHIS (US Department of Agriculture Animal and Plant Health Inspection Service) (2000b) Irradiation phytosanitary treatment of imported fruits and vegetables. *Federal Register* 65, 34113–34125.

APHIS (US Department of Agriculture Animal and Plant Health Inspection Service) (2002a) Importation of clementines from Spain. *Federal Register* 67, 45922–45933.

APHIS (US Department of Agriculture Animal and Plant Health Inspection Service) (2002b) Irradiation phytosanitary treatment of imported fruit and vegetables (final rule). *Federal Register* 67, 65016–65029.

APHIS (US Department of Agriculture Animal and Plant Health Inspection Service) (2007) Treatment Manual. Available at: http://www.aphis.usda.gov/ppq/manuals/port/Treatment_Chapters.htm (accessed 25 June 2007).

APPPC (Asia and Pacific Plant Protection Commission) (2005) *Guidelines for the Confirmation of Non-host Status of Fruit and Vegetables to Tephritid Fruit Flies.* Regional Standards for Phytosanitary Measures No. 4. Regional Office for Asia and the Pacific (RAP) Publication 2005/27. APPPC, Bangkok, Thailand.

APVMA (Australian Pesticides and Veterinary Medicines Authority) (2007) Chemical Review Program and Reports. Available at: http://www.apvma.gov.au/chemrev/chemrev.shtml (accessed 25 June 2007).

AQIS (Australian Quarantine Inspection Service) (2003) Import Risk Analysis Handbook 2003. Available at: http://www.daff.gov.au/content/publications.cfm?objectid=d667dce6-a412-4673-a6b49b7579cf4ad7 (accessed 25 June 2007).

Armstrong, J.W. (1982) Development of a hot-water immersion quarantine treatment for Hawaiian-grown Brazilian bananas. *Journal of Economic Entomology* 75, 787–790.

Armstrong, J.W. (1991) 'Sharwil' avocado: quarantine security against fruit fly (Diptera: Tephritidae) infestation in Hawaii. *Journal of Economic Entomology* 84, 1308–1315.

Armstrong, J.W. (1994) Commodity resistance to infestation by quarantine pests. In: Sharp, J.L. and Hallman, G.J. (eds) *Quarantine Treatments for Pests of Food Plants.* Westview Press, Boulder, Colorado, pp. 199–211.

Armstrong, J.W. (2001) Quarantine security of bananas at harvest maturity against Mediterranean and oriental fruit flies (Diptera: Tephritidae) in Hawaii. *Journal of Economic Entomology* 94, 302–314.

Armstrong, J.W. and Couey, H.M. (1989) Fruit disinfestation: fumigation, heat and cold. In: Robinson, A.S. and Hooper, G.H.S. (eds) *World Crop Pests.* Volume 3B: *Fruit Flies, their Biology, Natural Enemies and Control.* Elsevier, Amsterdam, pp. 411–424.

Armstrong, J.W. and Whitehand, L.C. (2005) Effects of methyl bromide fumigation, and fumigation temperature on Mediterranean fruit fly (Diptera: Trypetidae) egg and larval survival. *Journal of Economic Entomology* 98, 1116–1123.

Armstrong, J.W., Hansen, J.D., Hu, B.K.S. and Brown, S.A. (1989) High temperature forced-air quarantine treatment for papayas infested with tephritid fruit flies (Diptera: Tephritidae). *Journal of Economic Entomology* 82, 1667–1674.

Armstrong, J.W., Silva, S.T. and Shishido, V.M. (1995) Quarantine cold treatment for Hawaiian carambola fruit infested with Mediterranean fruit fly, melon fly, or oriental fruit fly (Diptera: Tephritidae) eggs and larvae. *Journal of Economic Entomology* 88, 683–687.

Arthur, V., Caceres, C., Wiendl, F.M. and Wiendl, J.A. (1993a) Controle da infestação natural de *Ceratitis capitata* (Wied., 1824) (Diptera, Tephritidae) em pêssegos (*Prunus persica*) através das radiações gama. *Scientia Agricola, Piracicaba* 50, 329–332.

Arthur, V., Wiendl, F.M. and Wiendl, J.A. (1993b) Controle de *Ceratitis capitata* (Wied., 1824) (Diptera, Tephritidae) em pêssegos (*Prunus persica*) infestados artificialmente através da radiação gama do cobalto-60. *Revue de Agricultura, Piracicaba* 68, 323–330.

ASTM International (2007) Guide for Irradiation for Fresh Agricultural Produce as a Phytosanitary Treatment. F 1355-06 Standard. Available at: http://www.astm.org/ (accessed 25 June 2007).

Aung, L.H., Leesch, J.G., Jenner, J.F. and Grafton-Cardwell, E.E. (2001) Effects of carbonyl sulfide, methyl iodide, and sulfuryl fluoride on fruit toxicity and insect mortality. *Annals of Applied Biology* 139, 93–100.

Aung, L.H., Leesch, J.G. and Jenner, J.F. (2004) Methyl iodide and forced aeration on the postharvest quality of lemons. *Postharvest Biology and Technology* 33, 45–50.

Back, E.A. and Pemberton, C.E. (1916) Effect of cold-storage temperatures upon the Mediterranean fruit fly. *Journal of Agricultural Research* 5, 657–666.

Bailey, S.W. (1962) The effects of percussion on insect pests of grain. *Journal of Economic Entomology* 55, 301–304.

Baker, A.C. (1939) *The Basis for Treatment of Products where Fruit Flies are Involved as a Condition for Entry to the USA.* Circular No. 551. US Department of Agriculture, Washington, DC.

Baker, R.H.A. (2002) Predicting the limits to the potential distribution of alien crop pests. In: Hallman, G.J. and Schwalbe, C.P. (eds) *Invasive Arthropods in Agriculture: Problems and Solutions.* Science Publishers, Enfield, North Carolina, pp. 207–241.

Baker, R.H.A., Sansford, C.E., Jarvis, C.H., Cannon, R.J.C., MacLeod, A. and Walters, K.F.A. (2000) The role of climatic mapping in predicting the potential geographical distribution of non-indigenous pests under current and future climates. *Agriculture, Ecosystems and Environment* 82, 57–71.

Baker, R.T., Cowley, J.M., Harte, D.S. and Frampton, E.R. (1990) Development of a maximum pest limit for fruit flies (Diptera: Tephritidae) in produce imported into New Zealand. *Journal of Economic Entomology* 83, 13–17.

Bakri, A., Heather, N., Hendrichs, J. and Ferris, I. (2005) Fifty years of radiation biology in entomology: lessons learned from IDIDAS. *Annals of the Entomological Society of America* 98, 1–12.

Banks, H.J. (1994) Fumigation – an endangered technology. In: Highley, E., Wright, E.J., Banks, H.J. and Champ, B.R. (eds) *Stored Product Protection*. Vol. 1. CAB International, Wallingford, UK, pp. 2–6.

Banks, H.J., Annis, P.C. and Rigby, G.R. (1990) Controlled atmosphere storage of grain. In: Fleurat-Lessard, F. and Ducom, P. (eds) *Proceedings of the Fifth International Working Conference on Stored-product Protection*. Bordeaux, France, pp. 695–706.

Bansiddhi, K., Siriphontangmun, S., Rumchiapikul, S., Kiennmeesuke, P., Prasongsab, S., Jaipetch, U., Kasa, S., Onboon, P. and Kamnungsak, V. (2004) Combination treatments with irradiation for controlling orchid thrips, *Thrips palmi*. *Irradiation as a Phytosanitary Treatment of Food and Agricultural Commodities*. IAEA-TECDOC-1427. International Atomic Energy Agency, Vienna, pp. 155–161.

Barsotti, L. and Cheftel, J.C. (1999) Traitement des aliments par champs électriques pulsés. II. Aspects biologiques. *Sciences des aliments* 19, 3–33.

Bartlett, A.C. (1985) Guidelines for genetic diversity in laboratory colony establishment and maintenance. In: Singh, P. and Moore, R.F. (eds) *Handbook of Insect Rearing*. Vol. I. Elsevier, Amsterdam, pp. 7–17.

Bateman, M.A. (1967) Adaptations to temperature in geographic races of the Queensland fruit fly, *Dacus* (*Strumeta*) *tryoni*. *Australian Journal of Zoology* 15, 1141–1161.

Bateman, M.A. (1982) Chemical methods for suppression or eradication of fruit fly populations. In: Drew, R.A.I., Hooper, G.H.S. and Bateman, M.A. (eds) *Economic Fruit Flies of the South Pacific Region*. Queensland Department of Primary Industries, Brisbane, Australia, pp. 115–128.

Bateman, M.A. and Morton, T.C. (1981) The importance of ammonia in proteinaceous attractants for fruit flies (Family: Tephritidae). *Australian Journal of Agricultural Research* 32, 883–903.

Beavis, C., Simpson, P., Syme, J. and Ryan, C. (1991) *Chemicals for the Protection of Fruit and Nut Crops*. Department of Primary Industries, Queensland Government, Brisbane, Australia.

Bengston, M. (1976) Timed daily emission of dichlorvos for control of *Ephestia cautella* infesting stored wheat. *Journal of Stored Product Research* 12, 157–164.

Benschoter, C.A. (1984) Low-temperature storage as a quarantine treatment for the Caribbean fruit fly (Diptera: Tephritidae) in Florida citrus. *Journal of Economic Entomology* 77, 1233–1235.

Benschoter, C.A. (1987) Effects of modified atmospheres and refrigeration on survival of eggs and larvae of Caribbean fruit fly (Diptera: Tephritidae) in laboratory diet. *Journal of Economic Entomology* 80, 1223–1225.

Benschoter, C.A., Spalding, D.H. and Reeder, W.F. (1981) Toxicity of atmospheric gases to immature stages of *Anastrepha suspensa*. *Florida Entomologist* 64, 543–544.

Birla, S.L., Wang, S., Tang, J. and Hallman, G. (2004) Improving heating uniformity of fresh fruit in radio frequency treatments for pest control. *Postharvest Biology and Technology* 33, 205–217.

Boag, T.S., Johnson, G.I., Izard, M., Murray, C. and Fitzsimmons, K.C. (1990) Physiological responses of mangoes cv. Kensington Pride to gamma irradiation treatment as affected by fruit maturity and ripeness. *Annals of Applied Biology* 116, 177–187.

Bollen, A.K. and Dela Rue, B.T. (1999) Hydrodynamic heat transfer – a technique for disinfestation. *Postharvest Biology and Technology* 17, 133–141.

Bond, E.J. (1984) *Manual of Fumigation for Insect Control*, 3rd edn. Food and Agriculture Organization of the United Nations, Rome.

Braithwaite, R.M. (1963) The sterilisation of banana fruit against Queensland fruit fly (*Strumeta tryoni*). *Australian Journal of Experimental Agriculture and Animal Husbandry* 3, 98–100.

Brash, D., Epenhuijsen, K. van, Zhang, Z. and Carpenter, A. (2002) Fumigating to control thrips. *Grower* 57, 24–25.

Bures, B.L., Donohue, K.V., Roe, R.M. and Bourham, M.A. (2005) Visualization of helium dielectric barrier discharge treatment of green peach aphids on tobacco leaves. *Transactions on Plasma Sciences* 33, 290–291.

Bures, B.L., Donohue, K.V., Roe, R.M. and Bourham, M.A. (2006) Nonchemical dielectric barrier discharge treatment as a method of insect control. *Transactions on Plasma Sciences* 34, 55–62.

Burg, S.P. (2004) *Postharvest Physiology and Hypobaric Storage of Fresh Produce.* CABI, Wallingford, UK.

Burkholder, W.E. and Ma, M. (1985) Pheromones for monitoring and control of stored-product insects. *Annual Review of Entomology* 30, 237–272.

Burkholder, W.E., Tilton, E.W. and Cogburn, R.R. (1966) Effects of gamma radiation on the grain mite *Acarus siro. Journal Economic Entomology* 59, 976–980.

Bustos, M.E., Enkerlin, W., Reyes, J. and Toledo, J. (2004) Irradiation of mangoes as a postharvest quarantine treatment for fruit flies (Diptera: Tephritidae). *Journal of Economic Entomology* 97, 286–292.

Calderon, M. (1990) Introduction. In: Calderon, M. and Barkai-Golan, R. (eds) *Food Preservation by Modified Atmospheres.* CRC Press, Boca Raton, Florida, pp. 3–8.

Calderon, M. and Barkai-Golan, R. (eds) (1990) *Food Preservation by Modified Atmospheres.* CRC Press, Boca Raton, Florida.

Calvitti, M., Govini, C., Buttarazzi, M. and Cirio, U. (1997) Induced sterility in greenhouse whitefly (Homoptera: Aleyrodidae) treated with gamma radiation. *Journal of Economic Entomology* 90, 1022–1027.

Cameron, A. (2002) FREECALC Version 2. AusVet Animal Health Services. Available at: http://www.ausvet.com.au/content.php?page=res_software (accessed 25 June 2007).

Cameron, G., Pate, J. and Vogel, K.M. (2001) Planting fear: how real is the threat of agricultural terrorism? *Bulletin Atomic Scientists* 57, 38–44.

Carey, J.R. (1991) Establishment of the Mediterranean fruit fly in California. *Science* 253, 1369–1373.

Carpenter, A. (1995) Implementation of controlled atmosphere disinfestation of export asparagus. *Proceedings of the 48th New Zealand Plant Protection Conference 1995,* 318–321.

Carpenter, A. and Potter, M. (1994) Controlled atmospheres. In: Sharp, J.L. and Hallman, G.J. (eds) *Quarantine Treatments for Pests of Food Plants.* Westview Press, Boulder, Colorado, pp. 171–198.

Casagrande, R. (2000) Biological terrorism targeted at agriculture: the threat to US national security. *Nonproliferation Review* 7, 92–105.

Castro, D., Espinosa, J. and Vargas, M. (2004) Ionizing radiation as a quarantine treatment for controlling *Brevipalpus chilensis* (Acarina: Tenuipalpidae) in Thompson seedless grapes. *Irradiation as a Phytosanitary Treatment of Food and Agricultural Commodities.* IAEA-TECDOC-1427. International Atomic Energy Agency, Vienna, pp. 143–153.

CBW (Convention on Biological Warfare) (2005) The Biological and Toxic Weapons web site. Convention on Biological Warfare. Available at: http:// opbw.org/ (accessed 25 June 2007).

CDFA (California Department of Food and Agriculture) (2007) Plant Quarantine Manual. Available at: http://pi.cdfa.ca.gov/pqm/manual/htm/pqm_index.htm (accessed 25 June 2007).

Chambers, J. (1990) Overview on stored-product insect pheromones and food attractants. *Journal of the Kansas Entomological Society* 63, 490–499.

Chapman, R.F. (1998) *The Insects: Structure and Function.* Cambridge University Press, Cambridge, UK.

COA (Council of Agriculture) (2006) Quarantine Requirements for the Importation of Plants or Plant Products into the Republic of China. Procedures for recognition of pest free areas. Available at: www.baphiq.gov.tw/main/object/images/pq/Pest%20 Free%20Areas.doc (accessed 25 June 2007).

Cogburn, R.R., Tilton, E.W. and Brower, J.H. (1973) Almond moth: gamma radiation effects on the life stages. *Journal of Economic Entomology* 66, 745–751.

Cohen, A.C. (2004) *Insect Diets: Science and Technology*. CRC Press, Boca Raton, Florida.

Corcoran, R.J. (2002) Fruit fly (Diptera: Tephritidae) responses to quarantine heat treatment. PhD thesis, The University of Queensland, Brisbane, Australia.

Corcoran, R.J., Heather, N.W. and Heard, T.A. (1993) Vapor heat treatment for zucchini infested with *Bactrocera cucumis* (Diptera: Tephritidae). *Journal of Economic Entomology* 86, 66–69.

Couey, H.M. and Chew, V. (1986) Confidence limits and sample size in quarantine research. *Journal of Economic Entomology* 79, 887–890.

Couey, H.M. and Hayes, C.F. (1986) Quarantine procedure for Hawaiian papaya using fruit selection and a two-stage hot-water immersion. *Journal of Economic Entomology* 79, 1307–1314.

CSIRO (Commonwealth Scientific and Industrial Research Organization) Entomology (2000) *Patent Portfolio*. CSIRO, Canberra, Australian Capital Territory, Australia.

Cunningham, I.C. (1991) Mango seed weevil in Queensland. *Acta Horticulturae* 291, 413–417.

Cunningham, R.T. (1989a) Parapheromones. In: Robinson, A.S. and Hooper, G. (eds) *Fruit Flies, their Biology, Natural Enemies and Control*. Vol. 3A. Elsevier, Amsterdam, The Netherlands, pp. 221–230.

Cunningham, R.T. (1989b) Population detection. In: Robinson, A.S. and Hooper, G. (eds) *Fruit Flies, their Biology, Natural Enemies and Control*. Vol. 3B. Elsevier, Amsterdam, The Netherlands, pp.169–173.

Davenport, T.L., Burg, S.P. and White, T.L. (2006) Optimal low-pressure conditions for long-term storage of fresh commodities kill Caribbean fruit fly eggs and larvae. *HortTechnology* 16, 98–104.

De Lima, C.P.F., Jessup, A.J., Cruickshank, L., Walsh, C.J. and Mansfield, E.R. (2007) Cold disinfestation of citrus (*Citrus* spp.) for Mediterranean fruit fly (*Ceratitis capitata*) and Queensland fruit fly (*Bactrocera tryoni*) (Diptera: Tephritidae). *New Zealand Journal of Crop and Horticultural Science* 35, 39–50.

Denlinger, D.L. and Lee, R.E. Jr (1998) Physiology of cold sensitivity. In: Hallman, G.J. and Denlinger, D.L. (eds) *Thermal Sensitivity in Insects and Application in Integrated Pest Management*. Westview Press, Boulder, Colorado, pp. 55–95.

Denlinger, D.L. and Yocum, G.D. (1998) Physiology of heat sensitivity. In: Hallman, G.J. and Denlinger, D.L. (eds) *Thermal Sensitivity in Insects and Application in Integrated Pest Management*. Westview Press, Boulder, Colorado, pp. 8–53.

Dentener, P.R., Waddell, B.C. and Batchelor, T.A. (1990) Disinfestation of lightbrown apple moth: a discussion of three methods. In: Beattie, B.B. (ed.) *Managing Postharvest Horticulture in Australia*. Australian Institute of Agricultural Science Occasional Publication No. 46. Australian Institute of Agricultural Science, Melbourne, Victoria, Australia, pp. 166–177.

Dentener, P.R., Peetz, S.M. and Birtles, D.B. (1992) Modified atmospheres for the postharvest disinfestation of New Zealand persimmons. *New Zealand Journal of Crop and Horticultural Sciences* 20, 203–208.

De Pietri-Tonelli, P. and Barontini, A. (1957) Activity of N-monomethylamide of 0, O-dimethyl dithiophospholylactic acid (L395), of parathion and their mixture with copper sulphate against *Dacus oleae*. *Olivicoltura* 12, 6 (from *Chemical Abstracts* 1957, 17060i).

Dermott, T. and Evans, D.E. (1977) An evaluation of fluidized bed heating as a means of disinfesting wheat. *Journal of Stored Product Research* 14, 1–12.

Desmarchelier, J.M. (1994) Carbonyl sulphide as a fumigant for control of insects and mites. In: Highley, E., Wright, E.J., Banks, H.J. and Champ, B.R. (eds) *Stored Product Protection*. Vol. 1. CAB International, Wallingford, UK, pp. 78–82.

Devorshak, C. and Griffin, R. (2002) Role and relationship of official and scientific information concerning pest status. In: Hallman, G.J. and Schwalbe, C.P. (eds) *Invasive Arthropods in Agriculture: Problems and Solutions*. Science Publishers, Enfield, New Hampshire, pp. 51–70.

Dohino, T. and Masaki, S. (1995) Effects of electron beam irradiation on Comstock mealy bug, *Pseudococcus comstocki* (Kuwana) (Homoptera: Pseudococcidae). *Research Bulletin of Plant Protection in Japan* 31, 31–36.

Dohino, T., Tanabe, K., Masaki, S. and Hayashi, T. (1996) Effects of electron beam irradiation on *Thrips palmi* Karny and *Thrips tabaci* Lindeman (Thysanoptera: Thripidae). *Research Bulletin of Plant Protection in Japan* 32, 23–29.

Donohue, K.V., Bures, B.L., Bourham, M.A. and Roe, R.M. (2006) Mode of action of a novel nonchemical method of insect control: atmospheric pressure plasma discharge. *Journal of Economic Entomology* 99, 38–47.

Drew, R.A.I. and Fay, H.A.C. (1988) Comparison of the roles of ammonia and bacteria in the attraction of *Dacus tryoni* (Froggatt) (Queensland fruit fly) to proteinaceous suspensions. *Journal of Plant Protection in the Tropics* 5, 127–130.

Drew, R.A.I. and Hancock, D.L. (1994) The *Bactrocera dorsalis* complex of fruit flies (Diptera: Tephritidae: Dacinae) in Asia. *Bulletin of Entomological Research Supplement Series*, Supplement No. 2, 68 pp.

Ducom, P. (2005) The issue of phosphine insecticide efficacy. In: Anon. (ed.) *Proceedings of the 2005 International Research Conference on Methyl Bromide Alternatives and Emission Reductions* 60, 1–3. Available at: www.mbao.org (accessed 25 June 2007).

Ducom, V. (1994) Methyl isothiocyanate used as a grain fumigant. In: Highley, E., Wright, E.J., Banks, H.J. and Champ, B.R. (eds) *Stored Product Protection*. Vol 1. CAB International, Wallingford, UK, pp. 91–97.

Dyck, V.A., Hendrichs, J. and Robinson, A.S. (eds) (2005) *Sterile Insect Technique: Principles and Practice in Area-wide Integrated Pest Management*. Springer, Heidelberg.

Ebbels, D.L. (2003) *Principles of Plant Health and Quarantine*. CAB International, Wallingford, UK.

Economopoulos, A.P. and Loukas, M.G. (1982) Selection at the alcohol dehydrogenase locus of the olive fruit fly under laboratory rearing: effect of larval diet and colony temperature. In: Cavalloro, R. (ed.) *Fruit Flies of Economic Importance*. Proceedings of the Commission of European Communities (CEC)/International Organisation for Biological and Integrated Control of Noxious Animals and Plants (IOBC) International Symposium, Athens, Greece, 16–19 November 1982, pp. 178–181.

EcoPort (2007) Ecoport. Available at: http://ecoport.org/ (accessed 25 June 2007).

Elith, J., Graham, C.H., Anderson, R.P., Dudík, M., Ferrier, S., Guisan, A., Hijmans, R.J., Huettmann, F., Leathwick, J.R., Lehmann, A., Li, J., Lohmann, L.G., Loiselle, B.A., Manion, G., Moritz, C., Nakamura, M., Nakazawa, Y., Overton, J.McC.M., Townsend-Peterson, A., Phillips, S.J., Richardson, K., Scachetti-Pereira, R., Schapire, R.E., Soberón, J., Williams, S., Wisz, M.S. and Zimmermann, N.E. (2006) Novel methods to improve prediction of species distributions from occurrence data. *Ecography* 29, 129–151.

EPA (Environmental Protection Agency) (1984) Rules and regulations. Revocation of tolerance for ethylene dibromide. *Federal Register* 49(103), 22082–22085.

EPA (Environmental Protection Agency) (2005a) Ozone Depletion Rules and Regulations. Available at: http://www.epa.gov/ozone/mbr/ (accessed 25 June 2007).

EPA (Environmental Protection Agency) (2005b) Sulphuryl fluoride; pesticide tolerance. *Federal Register* 70, 40899–40908.

EPA (Environmental Protection Agency) (2007) Regulatory Changes Proposed on Insecticide Pest Strip and Other DDVP Products. Available at: http://www.epa.gov/pesticides/op/ddvp/ddvp_changes.htm (accessed 25 June 2007).

Epenhuijsen, C.W. van, Koolaard, J.P. and Potter, J.F. (1997) Energy, ultrasound and chemical treatments for the disinfestation of fresh asparagus spears. *Proceedings of the 50th New Zealand Plant Protection Conference* 50, 436–441.

Epenhuijsen, C.W. van, Bodger, P.S., Koolaard, J.P., Woudberg, J.J. and Johnstone, P.T. (2001) Electromagnetic treatment of New Zealand flower thrips (*Thrips obscuratus*, Thysanoptera: Thripidae) in deionized water. *Journal of Microwave Power and Electromagnetic Energy* 36, 187–192.

Epenhuijsen, C.W. van, Carpenter, A. and Butler, R. (2002) Controlled atmospheres for the postharvest control of *Myzus persicae* (Sulzer) (Homoptera: Aphididae): effects of carbon dioxide concentration. *Journal of Stored Product Research* 38, 281–291.

EPPO (European and Mediterranean Plant Protection Organization) (1992) *Quarantine Pests for Europe*. CAB International, Wallingford, UK.

EPPO (European and Mediterranean Plant Protection Organization) (1997) Guidelines on Pest Risk Analysis: Pest Risk Assessment Scheme. Available at: http://www.eppo.org/QUARANTINE/quarantine.htm (accessed 25 June 2007).

EPPO (European and Mediterranean Plant Protection Organization) (2006) Guidelines on pest risk analysis: decision support scheme for quarantine pests. http://www.eppo.org/QUARANTINE/quaratine.htm (accessed 14 October 2007).

EPPO (European and Mediterranean Plant Protection Organization) (2007a) Data Sheets on Quarantine Pests. Available at: www.eppo.org/QUARANTINE/insects/Eutetranychus_orientalis/EUTEOR_ds.pdf (accessed 25 June 2007).

EPPO (European and Mediterranean Plant Protection Organization) (2007b) Phytosanitary procedures. http://archives.eppo.org/EPPOStandards/procedures.htm (accessed 25 June 2007).

Eustice, R.F. and Bruhn, C.M. (2006) Consumer acceptance and marketing of irradiated foods. In: Sommers, C.H. and Fan, X. (eds) *Food Irradiation Research and Technology*. Blackwell, Ames, Iowa, pp. 63–83.

Evans, D.E. (1981) Thermal acclimation in several species of stored-grain beetles. *Australian Journal of Zoology* 29, 483–492.

Evans, D.E. and Dermott, T. (1981) Dosage-mortality relationships for *Rhyzopertha dominica* (F.) (Coleoptera: Bostrychidae) exposed to heat in a fluidised bed. *Journal of Stored Products Research* 17, 53–64.

FAO (Food and Agriculture Organization) (1984) *International Plant Quarantine Treatment Manual*. FAO Plant Production and Protection paper No. 50. FAO, Rome.

FAO (Food and Agriculture Organization) (1997) Appendix A International Plant Protection Convention. Available at: http://www.fao.org/Legal/TREATIES/004t-e.htm (accessed 20 August 2007).

FAO/IAEA (Food and Agriculture Organization/International Atomic Energy Agency) (2007) Joint FAO/IAEA International Database on Insect Disinfestation and Sterilization. Available at: http://www-infocris.iaea.org/IDIDAS/start.htm (accessed 25 June 2007).

FAO/WHO (Food and Agriculture Organization/World Health Organization) (2007) Pesticide Residues in Food: Maximum Residue Limits; Extraneous Maximum Residue Limits. Available at: http://www.codexalimentarius.net/mrls/pestdes/jsp/pest_q-e.jsp (accessed 25 June 2007).

FAS (Foreign Agricultural Service, USA Department of Agriculture) (2007) International MRL database. Available at: http://mrldatabase.com/query.cfm (accessed 25 June 2007).

FDA (Food and Drug Administration) (2004) Irradiation in the production, processing and handling of food. *Federal Register* 69, 76844–76847.

Fésüs, I., Kádas, L. and Kálmán, B. (1981) Protection of oranges by gamma radiation against *Ceratitis capitata* Wied. *Acta Alimentaria* 10, 293–299.

Finney, D.J. (1971) *Probit Analysis*, 3rd edn. Cambridge University Press, Cambridge, UK.

Fleming, M.R., Hoover, K., Janowiak, J.J., Fang, Y., Wang, X., Liu, W., Wang, Y., Hang, X., Agrawal, D., Mastro, V.C., Lance, D.R., Shield, J.E. and Roy, R. (2003) Microwave irradiation of wood packing material to destroy the Asian longhorned beetle. *Forest Products Journal* 53, 46–52.

Fleurat-Lessard, F. (1990) Effect of modified atmospheres on insects and mites infesting stored products. In: Calderon, M. and Barkai-Golan, R. (eds) *Food Preservation by Modified Atmospheres.* CRC Press, Boca Raton, Florida, pp. 21–38.

Follett, P.A. (2001) Irradiation as a quarantine treatment for mango seed weevil. *Proceedings of the Hawaiian Entomological Society* 35, 40–45.

Follett, P.A. (2006a) Irradiation as a methyl bromide alternative for postharvest control of *Omphisa anastomosalis* (Lepidoptera: Pyralidae) and *Euscepes postfasciatus* and *Cylas formicarius elegantulus* (Coleoptera: Curculionidae) in sweet potatoes. *Journal of Economic Entomology* 99, 32–37.

Follett, P.A. (2006b) Irradiation as a phytosanitary treatment for *Aspidiotus destructor* (Homoptera: Diaspididae). *Journal of Economic Entomology* 99, 1138–1142.

Follett, P.A. (2006c) Irradiation as a phytosanitary treatment for white peach scale (Homoptera: Diaspididae). *Journal of Economic Entomology* 99, 1974–1978.

Follett, P.A. and Armstrong, J.W. (2004) Revised irradiation doses to control melon fly, Mediterranean fruit fly, and oriental fruit fly (Diptera: Tephritidae) and a generic dose for tephritid fruit flies. *Journal of Economic Entomology* 97, 1254–1262.

Follett, P.A. and Gabbard, Z. (2000) Effect of mango seed weevil (Coleoptera: Curculionidae) damage on mango seed viability in Hawaii. *Journal of Economic Entomology* 93, 1237–1240.

Follett, P.A. and Griffin, R.L. (2006) Irradiation as a phytosanitary treatment for fresh horticultural commodities: research and regulations. In: Sommers, C.H. and Fan, X. (eds) *Food Irradiation Research and Technology*. Blackwell Publications, Ames, Iowa, pp. 143–168.

Follett, P.A. and Hennessey, M.K. (2007) Confidence limits and sample size for determining nonhost status of fruits and vegetables to tephritid fruit flies as a quarantine measure. *Journal of Economic Entomology* 100, 251–257.

Follett, P.A. and Lower, R.A. (2000) Irradiation to ensure quarantine security for *Cryptophlebia* spp. (Lepidoptera: Tortricidae) in sapindaceous fruits from Hawaii. *Journal of Economic Entomology* 93, 1848–1854.

Follett, P.A. and McQuate, G.T. (2001) Accelerated development of quarantine treatments for insects on poor hosts. *Journal of Economic Entomology* 94, 1005–1011.

Follett, P.A. and Neven, L.G. (2006) Current trends in quarantine entomology. *Annual Review of Entomology* 51, 359–385.

Follett, P.A. and Sanxter, S.S. (2001) Hot water immersion to ensure quarantine security for *Cryptophlebia* spp. (Lepidoptera: Tortricidae) in lychee and longan exported from Hawaii. *Journal of Economic Entomology* 94, 1292–1295.

Fons, J.F. (1990) Quarantine barriers, treatments as a means to facilitate trade. *Acta Horticulturae* 269, 435–439.

Forney, C.F. and Houck, L.G. (1994) Chemical treatments. In: Paull, R.E. and Armstrong, J.W. (eds) *Insect Pests and Fresh Horticultural Products: Treatments and Responses.* CAB International, Wallingford, UK, pp. 139–162.

Franz, G., Gencheva, E. and Kerremans, P. (1994) Improved stability of genetic sex-separation strains for the Mediterranean fruit fly, *Ceratitis capitata. Genome* 37, 72–82.

Fricke, H. and Hart, E.J. (1967) Chemical dosimetry. In: Attix, F.H. and Roach, W.C. (eds) *Radiation Dosimetry 2.* Academic Press, New York, pp. 167–239.

FSANZ (Food Standards Australia and New Zealand) (2000) Proposal 22 – Review of Standard A17 – Irradiation of Food, and Standard A19 – Novel Foods. Available at: http://www.foodstandards.gov.au/standardsdevelopment/proposals/proposalp223reviewof1621.cfm. (accessed 25 June 2007).

Fuester, R.W., Kozempel, M.F., Forster, L.D., Goldberg, N., Casillas, L.I. and Swan, K.S. (2004) A novel nonchemical method for quarantine treatment of fruits: California red scale on citrus. *Journal of Economic Entomology* 97, 1861–1867.

Gaffney, J.J. and Armstrong, J.W. (1990) High temperature forced-air research facility for insect quarantine treatments. *Journal of Economic Entomology* 83, 1959–1964.

Gaffney, J.J., Hallman, G.J. and Sharp, J.L. (1990) Vapor heat research unit for insect quarantine treatments. *Journal of Economic Entomology* 83, 1965–1971.

Garrett, B.C. (1996) The Colorado potato beetle goes to war. *CBW Conventions Bulletin* 33, 2–3.

Gaunce, A.P., Morgan, C.V.G. and Meherhuik, M. (1982) Control of tree fruit insects with modified atmospheres. In: Richardson, D.G. and Meherhuik, M. (eds) *Controlled Atmospheres for Storage and Transport of Perishable Agricultural Commodities.* Oregon State University School of Agriculture, Symposium Series 1. Timber Press, Beaverton, Oregon, pp. 383–390.

Gold, C.S., Pinese, B. and Peña, J.E. (2002) Pests of bananas. In: Peña, J.E., Sharp, J.L. and Wysoki, M. (eds) *Tropical Fruit Pests and Pollinators.* CAB International, Wallingford, UK, pp. 13–56.

Goodwin, S. and Wellham, T.M. (1990) Gamma irradiation for disinfestation of cut flowers infested by twospotted spider mite (Acarina: Tetranychidae). *Journal of Economic Entomology* 83, 1455–1458.

Gould, W.P. (1994) Cold storage. In: Sharp, J.L. and Hallman, G.J. (eds) *Quarantine Treatments for Pests of Food Plants.* Westview Press Boulder, Colorado, pp. 119–132.

Gould, W.P. and Hallman, G.J. (2001) Laboratory and field infestation studies on monstera to determine its host status in relation to the Caribbean fruit fly (Diptera: Tephritidae). *Florida Entomologist* 84, 437–438.

Gould, W.P. and Hennessey, M.K. (1997) Mortality of *Anastrepha suspensa* (Diptera: Tephritidae) in carambolas treated with cold water precooling and cold storage. *Florida Entomologist* 80, 79–84.

Gould, W.P. and Sharp, J.L. (1992) Hot-water immersion quarantine treatment for guavas infested with Caribbean fruit fly (Diptera: Tephritidae). *Journal of Economic Entomology* 85, 1235–1239.

Gray, S.M. and Banerjee, N. (1999) Mechanisms of arthropod transmission of plant and animal viruses. *Microbiology and Molecular Biology Reviews* 63, 128–148.

Grode, J. and Christian, A. (2002) Regulatory enforcement programs within APHIS. In: Hallman, G.J. and Schwalbe, C.P. (eds) *Invasive Arthropods and Agriculture: Problems and Solutions.* Science Publishers, Enfield, New Hampshire, pp. 157–168.

Halfhill, J.E. (1988) Irradiation disinfestation of asparagus spears contaminated with *Brachycorynella asparagi* (Mordvilko) (Homoptera: Aphididae). *Journal of Economic Entomology* 81, 873–876.

Hallman, G.J. (1990) Vapor-heat treatment of carambolas infested with Caribbean fruit fly (Diptera: Tephritidae). *Journal of Economic Entomology* 83, 2340–2342.

Hallman, G.J. (1994a) Controlled atmospheres. In: Paull, R.E. and Armstrong, J.W. (eds) *Insect Pests and Fresh Horticultural Products: Treatments and Responses.* CAB International, Wallingford, UK, pp. 121–136.

Hallman, G.J. (1994b) Mortality of third-instar Caribbean fruit fly (Diptera: Tephritidae) reared at three temperatures and exposed to hot water immersion or cold storage. *Journal of Economic Entomology* 87, 405–408.

Hallman, G.J. (1996) Mortality of Caribbean fruit fly in coated fruits. In: McPheron, B.A. and Steck, G.J. (eds) *Fruit Fly Pests: A World Assessment of their Biology and Management.* St Lucie Press, Delray Beach, Florida, pp. 495–498.

Hallman, G.J. (1997) Mortality of Mexican fruit fly (Diptera: Tephritidae) immatures in coated grapefruits. *Florida Entomologist* 80, 324–328.

Hallman, G.J. (1998) Ionizing radiation quarantine treatments. *Annais da Sociedade Entomologica do Brasil* 27, 313–323.

Hallman, G.J. (1999) Ionizing radiation quarantine treatments against tephritid fruit flies. *Postharvest Biology and Technology* 16, 93–106.

Hallman, G.J. (2000) Expanding radiation quarantine treatments beyond fruit flies. *Agricultural and Forest Entomology* 2, 85–95.

Hallman, G.J. (2001a) Ionizing irradiation quarantine treatment against sweetpotato weevil (Coleoptera: Curculionidae). *Florida Entomologist* 84, 415–417.

Hallman, G.J. (2001b) Irradiation as a quarantine treatment. In: Molins, R. (ed.) *Food Irradiation: Principles and Applications.* Wiley Interscience, New York, pp. 113–130.

Hallman, G.J. (2002) Quarantine treatments: achieving market access for agricultural commodities in the presence of invasive species. In: Hallman, G.J. and Schwalbe, C.P. (eds) *Invasive Arthropods in Agriculture: Problems and Solutions.* Science Publishers, Enfield, New Hampshire, pp. 271–291.

Hallman, G.J. (2003) Ionizing irradiation quarantine treatment against plum curculio (Coleoptera: Curculionidae). *Journal of Economic Entomology* 96, 1399–1404.

Hallman, G.J. (2004a) Ionizing irradiation quarantine treatment against oriental fruit moth (Lepidoptera: Tortricidae) in ambient and hypoxic atmospheres. *Journal of Economic Entomology* 97, 824–827.

Hallman, G.J. (2004b) Irradiation disinfestation of apple maggot (Diptera: Tephritidae) in hypoxic and low-temperature storage. *Journal of Economic Entomology* 97, 1245–1248.

Hallman, G.J. (2005) Hypoxia reduces reproductive susceptibility of plum curculio (Coleoptera: Curculionidae) to ionizing radiation. *Florida Entomologist* 88, 208–210.

Hallman, G.J. (2006) *ISPM No. 18 Explanatory Document.* International Plant Protection Convention (IPPC) Available at: https://www.ippc.int/servlet/BinaryDownloaderServlet/119587_ISPM18_ED.pdf?filename=1143537627526_ISPM18_in_format.pdf&refID=119587 (accessed 29 October 2007).

Hallman, G.J. (2007) Phytosanitary measures to prevent the introduction of invasive species. In: Nentwig, W. (ed.) *Biological Invasions.* Springer, Berlin, pp. 367–384.

Hallman, G.J. and Armstrong, J.W. (1994) Heated air treatments. In: Sharp, J.L. and Hallman, G.J. (eds) *Quarantine Treatments for Pests of Food Plants.* Westview Press, Boulder, Colorado, pp. 149–163.

Hallman, G.J. and Chalot, D.S. (1993) Possible quarantine treatments for Florida agricultural food commodities. *Proceedings of the Florida State Horticultural Society* 106, 240–243.

Hallman, G.J. and Foos, J.F. (1996) Coating combined with dimethoate as a quarantine treatment against fruit flies (Diptera: Tephritidae). *Postharvest Biology and Technology* 7, 177–181.

Hallman, G.J. and Hellmich, R.L. (2007) Modified atmosphere storage may reduce efficacy of irradiation phytosanitary treatments. *Acta Horticulturae* (in press).

Hallman, G.J. and Loaharanu, P. (2002) Generic ionizing radiation quarantine treatments against fruit flies (Diptera: Tephritidae) proposed. *Journal of Economic Entomology* 95, 893–901.

Hallman, G.J. and Sharp, J.L. (1994) Radio frequency heat treatments. In: Sharp, J.L. and Hallman, G.J. (eds) *Quarantine Treatments for Pests of Food Plants.* Westview Press, Boulder, Colorado, pp. 165–170.

Hallman, G.J. and Worley, J.W. (1999) Gamma radiation doses to prevent adult emergence from immatures of Mexican and West Indian fruit flies (Diptera: Tephritidae). *Journal of Economic Entomology* 92, 967–973.

Hallman, G.J. and Zhang, Q.H. (1997) Inhibition of fruit fly (Diptera: Tephritidae) development by pulsed electric field. *Florida Entomologist* 80, 239–248.

Hallman, G.J., Niperos-Carriedo, M.O., Baldwin, E.A. and Campbell, C.A. (1994) Mortality of Caribbean fruit fly (Diptera: Tephritidae) immatures in coated fruits. *Journal of Economic Entomology* 87, 752–757.

Hamilton, D. (1988) Setting MRL values for pesticides in foods. *Queensland Agricultural Journal* 114, 225–228.

Hanboonsong, Y., Choosai, C., Panyiom, S. and Damak, S. (2002) Transovarial transmission of sugar cane white leaf phytoplasma in the insect vector *Matsumuratettix hiroglyphicus* (Matsumura). *Insect Molecular Biology* 11, 97–103.

Hansen, J.D. (2001) Ultrasound treatments to control surface pests. *HortTechnology* 11, 186–188.

Hansen, J.D. and Hara, A.H. (1994) A review of postharvest disinfestation of cut flowers and foliage with special reference to tropicals. *Postharvest Biology and Technology* 4, 193–212.

Hansen, J.D., Chan, H.T. Jr, Hara, A.H. and Tenbrink, V.L. (1991a) Phytotoxic reaction of Hawaiian cut flowers and foliage to hydrogen cyanide fumigation. *HortScience* 26, 53–56.

Hansen, J.D., Hara, A.H., Chan, H.T. Jr and Tenbrink, V.L. (1991b) Efficacy of hydrogen cyanide fumigation as a treatment for pests of Hawaiian cut flowers and foliage after harvest. *Journal of Economic Entomology* 84, 532–536.

Hansen, J.D., Heidt, M.L., Neven, L.G., Mielke, E.A., Bai, J., Chen, P.M. and Spotts, R.A. (2006) Effect of high-pressure hot water washing treatment on fruit quality, insects, and diseases in apples and pears. Part III: use of silicone-based materials and mechanical methods to eliminate surface pests. *Postharvest Biology and Technology* 40, 221–229.

Hara, A.H. and Jacobsen, C.M. (2005) Hot water immersion for surface disinfestation of *Maconellicoccus hirsutus* (Homoptera: Pseudococcidae). *Journal of Economic Entomology* 98, 284–288.

Hara, A.H., Hata, T.Y., Hu, B.K.S. and Tenbrink, V.L. (1993) Hot-water immersion as a potential quarantine treatment against *Pseudaulacaspis cockerelli* (Homoptera: Diaspididae). *Journal of Economic Entomology* 86, 1167–1170.

Hara, A.H., Yalemar, J.A., Jang, E.B. and Moy, J.H. (2002) Irradiation as a possible quarantine treatment for green scale *Coccus viridis* (Green) (Homoptera: Coccidae). *Postharvest Biology and Technology* 25, 349–358.

Hardenburg, R.E., Watada, E.A. and Wang, C.Y. (1986) *The Commercial Storage of Fruits, Vegetables, and Florist and Nursery Stocks.* US Department of Agriculture (USDA), Agricultural Research Service Handbook No. 66. USDA, Washington, DC.

Harte, D.S., Cowley, J.M. and Baker, R.T. (1995) Accounting for variability of naturally infested fruit used in disinfestation treatment efficacy trials. *Journal of Economic Entomology* 88, 441–446.

Hata, T.V., Hara, A.H., Hu, B.E.S., Kaneko, R.T. and Tenbrink, V.L. (1993) Field sprays and insecticidal dips after harvest for pest management of *Frankliniella occidentalis* and *Thrips palmi* (Thysanoptera: Thripidae) on orchids. *Journal of Economic Entomology* 86, 1483–1489.

Hatton, T.T., Cubbedge, R.H., von Windeguth, D.L. and Spalding, D.H. (1982) Phosphine as a fumigant for grapefruit infested by Caribbean fruit fly larvae. *Proceedings of the Florida State Horticultural Society* 95, 221–224.

Hayashi, T., Todoriki, S. and Nakakita, H. (1999) Effectiveness of electron irradiation as a quarantine treatment of cut flowers. *Irradiation as a Quarantine Treatment of Arthropod Pests.* IAEA-TECDOC-1082. International Atomic Energy Agency, Vienna, pp. 49–55.

Hayashi, T., Imamura, T., Todoriki, S.S., Miyanoshita, A. and Nakakita, H. (2004) Soft-electron treatment as a phytosanitary measure for stored-product pests. *Irradiation as a Phytosanitary Treatment of Food and Agricultural Commodities.* IAEA-TECDOC-1427. International Atomic Energy Agency, Vienna, pp. 67–73.

Hayes, C.F. (1994) Modelling heat and cold transfer. In: Paull, R.E. and Armstrong, J.W. (eds) *Insect Pests and Fresh Horticultural Products: Treatments and Responses.* CAB International, Wallingford, UK, pp. 237–248.

Heather, N.W. (1975) West Indian drywood termite in Queensland. *Queensland Agricultural Journal* 101, 91–94.

Heather, N.W. (1994) Pesticide quarantine treatments. In: Sharp, J.L. and Hallman, G.J. (eds) *Quarantine Treatments for Pests of Food Plants.* Westview Press, Boulder, Colorado, pp. 89–100.

Heather, N.W., Hargreaves, P.A., Corcoran, R.J. and Melksham, K.J. (1987) Dimethoate and fenthion as packing line treatments for tomatoes against *Dacus tryoni*. *Australian Journal of Experimental Agriculture* 27, 465–469.

Heather, N.W., Walpole, D.E., Corcoran, R.J., Hargreaves, P.A. and Jordan, R.A. (1992) Postharvest quarantine disinfestation of zucchinis and rockmelons against *Bactrocera cucumis* (French) using insecticide dips of fenthion or dimethoate. *Australian Journal of Experimental Agriculture* 32, 241–244.

Heather, N.W., Corcoran, R.J. and Kopittke, R.A. (1997) Hot air disinfestation of Australian 'Kensington' mangoes against two fruit flies (Diptera: Tephritidae). *Postharvest Biology and Technology* 10, 99–105.

Heather, N.W., Lescano, H.G. and Congdon, B.C. (1999) An irradiation marker for mango seed weevil. *Irradiation as a Quarantine Treatment of Arthropod Pests.* IAEA-TECDOC-1082. International Atomic Energy Agency, Vienna, pp. 163–167.

Heather, N.W., Kopittke, R.A. and Pike, E.A. (2002) A heated air quarantine disinfestation treatment against Queensland fruit fly (Diptera: Tephritidae) for tomatoes. *Australian Journal of Experimental Agriculture* 42, 1125–1129.

Highley, E., Wright, E.J. and Champ, B.R. (eds) (1994) *Stored Product Protection.* Vol. 2. CAB International, Wallingford, UK.

Holt, J., Black, R. and Abdallah, R. (2006) A rigorous yet simple quantitative risk assessment method for quarantine pests and non-native organisms. *Annals of Applied Biology* 149, 167–173.

Horn, F., Horn, P. and Sullivan, J. (2005a) Current practice in fresh fruit fumigation with phosphine in Chile. In: Anon. (ed.) *Proceedings of the 2005 International Research Conferences on Methyl Bromide Alternatives and Emission Reductions* 61, 1–3. Available at: www.mbao.org (accessed 25 June 2007).

Horn, J., Horn, P., Horn, F. and Sullivan, J. (2005b) Control of false Chilean mite (*Brevipalpus chilensis*) with phosphine and cold storage. In: Anon. (ed.) *Proceedings of the 2005 International Research Conference on Methyl Bromide Alternatives and Emission Reductions* 62, 1–3. Available at: www.mbao.org (accessed 25 June 2007).

Hostetler, K. (2002) Purdue professor responds to accusations of US agroterrorism. *Purdue News.* Available at: http://www.purdue.edu/UNS/html4ever/021210.Edwards.rootworm.html (accessed 25 June 2007).

IAEA (International Atomic Energy Agency) (2001) *Standardised Methods to Verify Absorbed Dose in Irradiated Food for Insect Control.* IAEA-TECDOC-1201. International Atomic Energy Agency, Vienna, 124 pp.

ICGFI (International Consultative Group on Food Irradiation) (1991) *Code of Good Irradiation Practice for the Control of Pathogens and other Microflora in Spices, Herbs and other Vegetable Seasonings.* ICGFI Document No. 5. Food and Agriculture Organization/International Atomic Energy Agency (FAO/IAEA), Vienna.

ICGFI (International Consultative Group on Food Irradiation) (1992) *Irradiation as a Quarantine Treatment of Fresh Fruits and Vegetables.* ICGFI Document No. 13. Food and Agriculture Organization/International Atomic Energy Agency (FAO/IAEA), Vienna.

ICGFI (International Consultative Group on Food Irradiation) (1994) *Irradiation as a Quarantine Treatment of Fresh Fruits and Vegetables.* ICGFI Document No. 17. Food and Agriculture Organization/International Atomic Energy Agency (FAO/IAEA), Vienna.

Ignatowicz, S. (1999) Detection methods for mites and insects. *Irradiation as a Quarantine Treatment of Arthropod Pests.* IAEA-TECDOC-1082. International Atomic Energy Agency, Vienna, pp. 141–161.

Ignatowicz, S. (2004) Irradiation as an alternative to methyl bromide fumigation of agricultural commodities infested with quarantine stored-product pests. *Irradiation as a Phytosanitary Treatment of Food and Agricultural Commodities.* IAEA-TECDOC-1427. International Atomic Energy Agency, Vienna, pp. 51–66.

Ikediala, J.N., Hansen, J.D., Tang, J., Drake, S.R. and Wang, S. (2002) Development of a saline immersion technique with RF energy as a postharvest treatment against codling moth in cherries. *Postharvest Biology and Technology* 24, 209–221.

IPPC (International Plant Protection Convention) (2007) Standards (ISPM) International Phytosanitary Portal (IPP). Available at: https://www.ippc.int/IPP/En/default.jsp (accessed 25 June 2007).

IPSAPH (International Portal on Food Safety, Animal and Plant Health) (2007) Pesticide: Dichlorvos – Commodity: Mushrooms – MRL – 0.5 (mg/kg). Available at: www.ipfsaph.org/id/PESTMRL25 (accessed 25 June 2007).

Isikber, A.A., Navarro, S., Finkelman, S., Rindner, M. and Dias, R. (2004) Propylene oxide as a potential quarantine fumigant for insect disinfestation of nuts. In: Anon. (ed.) *Proceedings of 2004 International Research Conference on Methyl Bromide Alternatives and Emission Reductions* 52, 1–2. Available at: www.mbao.org (accessed 25 June 2007).

ISO (International Standards Organisation) (2002) Standard Guide for Dosimetry in Radiation Research on Food and Agricultural Products. ISO/ASTM51900–02. Available at: http://www.iso.ch/iso/en/CatalogueDetailPage.CatalogueDetail?CSNUMBER=33562&ICS1=67&ICS2=20&ICS3= (accessed 25 June 2007).

Iwata, M., Makiguchi, S., Ishikawa, A., Shimabukuro, S. and Tanabe, K. (1992) Acquisition of cold tolerance in immature stages of oriental fruit fly *Dacus dorsalis* (Diptera: Tephritidae) in artificial diets and orange fruits. *Research Bulletin Plant Protection Japan* 28, 55–60.

Jacobi, K., Coates, L. and Wong, L. (1994) Heat disinfestation of mangoes: effect on fruit quality and disease control. In: Champ, B.R., Highley, E. and Johnson, G.I. (eds) *Postharvest Handling of Tropical Fruits.* ACIAR (Australian Centre for International Agricultural Research) Proceedings No. 50. ACIAR, Canberra, Australia, pp. 280–287.

Jacobi, K., MacRae, E.A. and Hetherington, S.E. (2001) Effect of fruit maturity on the response of 'Kensington' mango fruit to heat treatment. *Australian Journal of Experimental Agriculture* 41, 793–803.

Jacobsen, C.M. and Hara, A.H. (2003) Irradiation of *Maconellicoccus hirsutus* (Homoptera: Pseudococcidae) for phytosanitation of agricultural commodities. *Journal of Economic Entomology* 96, 1334–1339.

Jang, E.B. (1990) Fruit fly disinfestation of tropical fruits using semipermeable shrinkwrap film. *Acta Horticulturae* 269, 453–458.

Jang, E.B. (1996) Systems approach to quarantine security: postharvest application of sequential mortality in the Hawaiian grown 'Sharwil' avocado system. *Journal of Economic Entomology* 89, 950–956.

Jang, E.B. and Moffitt, H.A. (1994) Systems approaches to achieving quarantine security. In: Sharp, J.L. and Hallman, G.J. (eds) *Quarantine Treatments for Pests of Food Plants.* Westview Press, Boulder, Colorado, pp. 225–237.

Jang, E.B., Nagata, J.T., Chan, H.T. Jr and Laidlaw, W.G. (1999) Thermal death kinetics in eggs and larvae of *Bactrocera latifrons* (Diptera: Tephritidae) and comparative thermotolerance to three other tephritid fruit fly species in Hawaii. *Journal of Economic Entomology* 92, 684–690.

Jessup, A.J., Sloggett, R.F. and Quinn, N.M. (1998) Quarantine disinfestation of blueberries against *Bactrocera tryoni* (Froggatt) (Diptera: Tephritidae) by cold storage. *Journal of Economic Entomology* 91, 964–967.

Johnson, J.A., Soderstrom, E.L., Brandl, D.G., Houck, L.G. and Wofford, P.A. (1990) Gamma radiation as a quarantine treatment for Fuller rose beetle eggs (Coleoptera: Curculionidae) on citrus fruit. *Journal of Economic Entomology* 83, 905–909.

Kader, A.A. and Ke, D. (1994) Controlled atmospheres. In: Paull, R.E. and Armstrong, J.W. (eds) *Insect Pests and Fresh Horticultural Products: Treatments and Responses.* CAB International, Wallingford, UK, pp. 223–236.

Kahn, R.P. (1979) A concept of pest risk analysis. *EPPO Bulletin* 9, 119–130.

Karnkowski, W. and Ignatowicz, S. (1999) Development of the potato cyst nematode *Globodera rostochieitsis* (Woll.) on potatoes after gamma radiation of cysts. *Irradiation as a Quarantine Treatment of Arthropod Pests*. IAEA-TECDOC-1082. International Atomic Energy Agency, Vienna, pp. 123–127.

Kelly, M.P. and Wilkin, D.R. (1994) The changing role of AQIS in the regulation of grain exports from Australia. In: Highley, E., Wright, E.J., Banks, H.J. and Champ, B.R. (eds) *Stored Product Protection*. Vol. 2. CAB International, Wallingford, UK, pp. 1186–1191.

Kikuchi, O.K., Wiendl, F.M. and Arthur, V. (1999) Tolerance of cut flowers to gamma-radiation. *Irradiation as a Quarantine Treatment of Arthropod Pests*. IAEA-TECDOC-1082. International Atomic Energy Agency, Vienna, pp. 93–104.

Koidsumi, K. (1930) Quantitative studies on the lethal action of X-rays upon certain insects. *Journal of the Society of Tropical Agriculture (Japan)* 2, 243–263.

Kopittke, R.A., Swain, A.J. and Wyatt, P.M. (2004) How many insects and what design should I use for quarantine research? *Proceedings of the International Congress of Entomology, Brisbane, Australia, 2004 (Abstracts)*. Electronic format.

Krishna, H., Ryan, R., Munez, A., Hirst, G., Yoshihara, H. and Barton, S. (2005) VAPORMATE™: disinfestation of Philippine export bananas. In: Anon. (ed.) *Proceedings of the 2005 International Research Conference on Methyl Bromide Alternatives and Emission Reductions* 119, 1–2. Available at: www.mbao.org (accessed 25 June 2007).

Krotchta, J.M., Baldwin, E.A. and Nisperos-Carriedo, M. (eds) (1994) *Edible Coatings and Films to Improve Food Quality*. Technomic, Lancaster, Pennsylvania.

Lagunas-Solar, M.C., Essert, T.K., Piña, U.C., Zeng, N.X. and Truong, T.D. (2006) Metabolic stress disinfection and disinfestation (MSDD): a new, non-thermal, residue-free process for fresh agricultural products. *Journal of the Science of Food and Agriculture* 86, 1814–1825.

Landolt, P.J., Chambers, D.L. and Chew, V. (1984) Alternative to use of probit 9 mortality as a criterion for quarantine treatments of fruit fly (Diptera: Tephritidae)-infested fruit. *Journal of Economic Entomology* 77, 285–287.

Laughlin, R. (1965) Capacity for increase: a useful population statistic. *Journal of Animal Ecology* 34, 77–91.

Lee, S. (1968) Effectiveness of various insecticides for destroying oriental fruit fly, *Dacus dorsalis* Hendel, infestations in green mature banana. *Journal of Taiwan Agricultural Research* 17, 17–25.

Leesch, J.G. (2002) Recapturing methyl bromide emissions. US Department of Agriculture (USDA) Agricultural Research Service (ARS) Products and Services. Available at: http://www.ars.usda.gov/is/np/mba/april96/leesch.htm (accessed 25 June 2007).

Leesch, J.G., Armstrong, J.W., Tebbets, J.S. and Tebbets, J.C. (2003) Insect control with ozone gas as an alternative to methyl bromide. In: Anon. (ed.) *Proceedings of 2003 International Research Conference on Methyl Bromide Alternatives and Emission Reductions* 63, 1–2. Available at: www.mbao.org (accessed 25 June 2007).

Lescano, H.G., Congdon, B.C. and Heather, N.W. (1994) Comparison of two potential methods to detect *Bactrocera tryoni* (Diptera: Tephritidae) gamma-irradiated for quarantine purposes. *Journal of Economic Entomology* 87, 1256–1261.

Lester, P.J. and Greenwood, D.R. (1997) Pretreatment induced thermotolerance in lightbrown apple moth (Lepidoptera: Tortricidae) and associated induction of heat shock protein synthesis. *Journal of Economic Entomology* 90, 199–204.

Lewis, V.R., Power, A.B. and Haverty, M.I. (2000) Laboratory evaluation of microwaves for control of the western drywood termite. *Forest Products Journal* 50, 79–87.

Liebhold, A.M., Work, T.T., McCullough, D.G. and Cavey, J.F. (2006) Airline baggage as a pathway for alien insect species invading the USA. *American Entomologist* 52, 48–54.

Liquido, N.J., Chan, H.T. and McQuate, G.T. (1995) Hawaiian tephritid fruit flies (Diptera): integrity of the infestation-free quarantine procedure for Sharwil avocado. *Journal of Economic Entomology* 88, 85–86.

Liquido, N.J., Barr, P.G. and Chew, V. (1997) CQT-STATS: Biological Statistics for Pest Risk Assessment in Developing Commodity Quarantine Treatment. US Department of Agriculture, Agricultural Research Service, Hilo, Hawaii.

Liu, Y.-B. (2003) Effects of vacuum and controlled atmosphere treatments on insect mortality and lettuce quality. *Journal of Economic Entomology* 96, 1100–1107.

Lockwood, J.A. (1987) Entomological warfare: history of the use of insects as weapons of war. *Bulletin of the Entomological Society of America* 33, 76–82.

Looney, R. (2002) Economic costs to the USA stemming from the 9/11 attacks. *Strategic Insights* 6, Center for Contemporary Conflict, US Navy. Available at: http://www.ccc.nps.navy.mil/si/aug02/homeland.asp (accessed 25 June 2007).

Lovell, B. (2006) Climate change: conflict of observational science, theory, and politics: discussion. *American Association of Petroleum Geologists Bulletin* 90, 405–407.

Low, A., Husing, U.P., Preisser, A. and Baur, X. (2003) Regulations and control of in-transit fumigated containers as well as of fumigated cargo ships. *International Maritime Health* 54, 77–85.

Lurie, S. (1998) Postharvest heat treatments of horticultural crops. *Horticultural Reviews* 22, 91–121.

Lurie, S. and Jang, E. (2007) The influence of heat shock proteins in insect pests and fruits in thermal treatments. In: Tang, J., Mitcham, E., Wang, S. and Lurie, S. (eds) *Heat Treatments for Postharvest Pest Control: Theory and Practice.* CAB International, Wallingford, UK, pp. 269–290.

Mack, R. and Barak, A. (2006) Alternative quarantine treatments for hitchhiking snails. In: Anon. (ed.) *Proceedings of 2006 International Research Conference on Methyl Bromide Alternatives and Emission Reductions* 146, 1. Available at: www.mbao.org (accessed 25 June 2007).

MacKenzie, D.R., Marchetti, M.A. and Kingsolver, C.H. (1985) The potential for the willful introduction of biotic agents as an act of anticrop warfare. In: MacKenzie, D.R., Barfield, C.S., Kennedy, G.G., Berger, R.D. and Taranto, D.J. (eds) *The Movement and Dispersal of Agriculturally Important Biotic Agents.* Claitor's Publishing Division, Baton Rouge, Louisiana, pp. 601–608.

Maddison, P.A. and Bartlett, B.J. (1989) A contribution towards the zoogeography of the Tephritidae. In: Robinson, A.S. and Hooper, G. (eds) *Fruit Flies, their Biology, Natural Enemies and Control.* Vol. 3A. Elsevier, Amsterdam, pp. 27–35.

Malavasi, A., Rohwer, G.G. and Campbell, D.S. (1994) Fruit fly free areas: strategies to develop them. In: Calkins, C.D., Klassen, W. and Liedo, P. (eds) *Fruit Flies and the Sterile Insect Technique.* CRC Press, Boca Raton, Florida, pp. 165–180.

Mangan, R.L. and Hallman, G.J. (1998) Temperature treatments for quarantine security: new approaches for fresh commodities. In: Hallman, G.J. and Denlinger, D.L. (eds) *Thermal Sensitivity in Insects and Application in Integrated Pest Management.* Westview Press, Boulder, Colorado, pp. 201–234.

Mangan, R.L. and Ingle, S.J. (1992) Forced hot-air quarantine treatment for mangoes infested with West Indian fruit fly (Diptera: Tephritidae). *Journal of Economic Entomology* 85, 1859–1864.

Mangan, R.L. and Sharp, J.L. (1994) Combination and multiple treatments. In: Sharp, J.L. and Hallman, G.J. (eds) *Quarantine Treatments for Pests of Food Plants.* Westview Press, Boulder, Colorado, pp. 239–247.

Mangan, R.L., Frampton, E.R., Thomas, D.T. and Moreno, D.S. (1997) Application of the maximum pest limit concept to quarantine security standards for the Mexican fruit fly (Diptera: Tephritidae). *Journal of Economic Entomology* 90, 1433–1440.

Mansour, M. (2003) Gamma irradiation as a quarantine treatment for apples infested by codling moth (Lepidoptera: Tortricidae). *Journal of Applied Entomology* 127, 137–141.

Mansour, M. and Franz, G. (1996) Gamma radiation as a quarantine treatment for the Mediterranean fruit fly (Diptera: Tephritidae). *Journal of Economic Entomology* 89, 1175–1180.

Matsumoto, G. (2003) Anthrax powder: state of the art? *Science* 302, 1492–1496.

May, A.W.S. (1953) Queensland host records for the Dacinae (fam. Trypetidae). *Queensland Journal of Agricultural Science* 10, 36–79.

May, A.W.S. (1962) The fruit fly problem in eastern Australia. *Journal of the Entomological Society of Queensland* 1, 1–4.

May, A.W.S. (1963) An investigation of fruit flies (Fam. Trypetidae) in Queensland. 1. Introduction, species, pest status and distribution. *Queensland Journal of Agricultural Science* 20, 1–82.

MBAO (Methyl Bromide Alternatives Outreach) (2007) Annual International Research Conferences on Methyl Bromide Alternatives and Emissions Reductions. Available at: www.mbao.org (accessed 25 June 2007).

Mbata, G.N. and Phillips, T.W. (2001) Effects of temperature and exposure time on mortality of stored-product insects exposed to low pressure. *Journal of Economic Entomology* 94, 1302–1307.

Mbata, G.N., Phillips, T.W. and Payton, M. (2004) Mortality of eggs of stored-product insects held under vacuum: effects of pressure, temperature, and exposure time. *Journal of Economic Entomology* 97, 695–702.

McCarthy, J.J., Canziani, O.F., Leary, N.A. and Dokken, D.J. (eds) (2001) *Climate Change 2001: Impacts, Adaptation, and Vulnerability.* Cambridge University Press, Cambridge, UK.

McDonald, R.E. and Miller, W.R. (1994) Quality and condition maintenance. In: Sharp, J.L. and Hallman, G.J. (eds) *Quarantine Treatments for Pests of Food Plants.* Westview Press, Boulder, Colorado, pp. 249–277.

McGuire, R.G. (1991) Concomitant decay reductions when mangoes are treated with heat to control infestations of Caribbean fruit flies. *Plant Disease* 75, 946–949.

McLauchlan, R.L., Barker, L.R. and Heather, N.W. (1993) Physiological aspects of low temperature disinfestation of mandarins. In: *Proceedings of the Australasian Postharvest Conference 1993.* The University of Queensland Gatton College, Lawes, Queensland, Australia, pp. 313–318.

Meats, A. (1976) Thresholds for cold-torpor and cold-survival in the Queensland fruit fly, and predictability of rates of change in survival threshold. *Insect Physiology* 22, 1506–1509.

Mitcham, E.J., Zhou, S. and Bikoba, V. (1997) Controlled atmospheres for quarantine control of three pests of table grape. *Journal of Economic Entomology* 90, 1360–1370.

Mitcham, E.J., Veltman, R.H., Feng, X., de Castro, E., Johnson, J.A., Simpson, T.L., Biasi, W.V., Wang, S. and Tang, J. (2004) Application of radio frequency treatments to control insects in in-shell walnuts. *Postharvest Biology and Technology* 33, 93–100.

Moffitt, H.R. (1971) Methyl bromide fumigation combined with cold storage for control of codling moth in apples. *Journal of Economic Entomology* 64, 1258–1260.

Moffitt, H.R. (1990) A systems approach to meeting quarantine requirements for insect pests of deciduous fruits. *Proceedings of the Washington State Horticultural Association* 85, 223–225.

Moffitt, H.R. and Albano, D.J. (1972) Effects of commercial fruit storage on stages of the codling moth. *Journal of Economic Entomology* 65, 770–773.

Molins, R. (ed.) (2001) *Food Irradiation: Principles and Applications.* Wiley Interscience, New York.

Moon, J.E.V.C. (2006) The US biological weapons program. In: Wheelis, M., Rózsa, L. and Dando, M. (eds) *Deadly Cultures: Biological Weapons since 1945.* Harvard University Press, Cambridge, Massachusetts, pp. 9–46.

Moss, J.L. and Chan, H.T. (1993) Thermal death kinetics of Caribbean fruit fly (Diptera: Tephritidae) embryos. *Journal of Economic Entomology* 86, 1162–1166.

Moy, J.H. and Wong, L. (2002) The efficacy and progress in using radiation as a quarantine treatment of tropical fruits – a case study in Hawaii. *Radiation Physics and Chemistry* 63, 397–401.

Moy, J.H., Chinnasri, B., Sipes, B.S., Schmitt, D.P., Hamasaki, R.T., Mersino, E.F. and Yamakawa, R.M. (1999) Radiation disinfestation of nematodes, aphids, mites, thrips and other pests on food plant materials: evaluation for effectiveness and product quality. *Irradiation as a Quarantine Treatment of Arthropod Pests*. IAEA-TECDOC-1082. International Atomic Energy Agency, Vienna, pp. 105–113.

Mumford, J.D. (2002) Economic issues related to quarantine in international trade. *European Review of Agricultural Economics* 29, 329–348.

Myburgh, A.C. and Bass, M.W. (1969) Effect of low temperature storage on pupae of false codling moth, *Cryptophlebia (Argyroploce) leucotreta* Meyr. *Phytophylactica* 1, 115–116.

Myers, J.H. and Hosking, G. (2002) Eradication. In: Hallman, G.J. and Schwalbe, C.P. (eds) *Invasive Arthropods in Agriculture: Problems and Solutions*. Science Publishers, Enfield, New Hampshire, pp. 293–308.

Nation, J.L., Smittle, B.J. and Milne, K. (1995) Radiation induced changes in melanization and phenoloxidase in Caribbean fruit fly larvae (Diptera: Tephritidae) as the basis of a simple test of irradiation. *Journal of Economic Entomology* 88, 201–205.

Nation, J.L., Smittle, B.J. and Milne, K. (1999) Physiological markers in insects indicating treatment with ionizing radiation. *Irradiation as a Quarantine Treatment of Arthropod Pests*. IAEA-TECDOC-1082. International Atomic Energy Agency, Vienna, pp. 129–139.

National Health and Medical Research Council (1988) *MRL Standard. Standard for Maximum Residue Limits of Pesticides, Agricultural Chemicals, Feed Additives, Veterinary Medicines and Noxious Substances in Food*. Australian Government Publishing Service, Canberra, Australian Capital Territory, Australia.

Navarro, S., Lider, O. and Gerson, U. (1985) Response of adults of the grain mite *Acarus siro* Linnaeus to modified atmospheres. *Journal of Agricultural Entomology* 2, 61–68.

Nel, R.G. (1936) The utilization of low temperatures in the sterilization of deciduous fruit infested with the immature stages of the Mediterranean fruit fly, *Ceratitis capitata* Wied. *Union of South Africa Department of Agriculture and Forestry Science Bulletin* 155.

Neven, L.G. (1998) Effects of heating rate on the mortality of fifth-instar codling moth (Lepidoptera: Tortricidae). *Journal of Economic Entomology* 91, 297–301.

Neven, L.G. (2003) Physiological effects of physical postharvest treatments on insects. *HortTechnology* 13, 272–275.

Neven, L.G. and Rehfield-Ray, L.M. (2006) Combined heat and controlled atmosphere quarantine treatments for control of western cherry fruit fly in sweet cherries. *Journal of Economic Entomology* 99, 653–658.

Neven, L.G., Hansen, J.D., Spotts, R.A., Serdani, M., Mielke, E.A., Bai, J., Chen, P.M. and Sanderson, P.G. (2006a) Effect of high-pressure hot water washing treatment on fruit quality, insects, and diseases in apples and pears. Part IV: use of silicone-based materials and mechanical methods to eliminate surface arthropod eggs. *Postharvest Biology and Technology* 40, 230–235.

Neven, L.G., Rehfield-Ray, L.M. and Obenland, D. (2006b) Confirmation and efficacy tests against codling moth and oriental fruit moth in peaches and nectarines using combination heat and controlled atmospheres. *Journal of Economic Entomology* 99, 1610–1619.

Nishijima, K.A., Chan, H.T. Jr, Sanxter, S.S. and Linse, E.S. (1995) Reduced heat shock period of 'Sharwil' avocado for cold tolerance in quarantine cold treatment. *HortScience* 30, 1052–1053.

Noble, A. (1983) Fruit and vegetables – stability of dimethoate in dips. *Refnote R6* Agdex/681. Queensland Department of Primary Industries, Brisbane, Australia.

NPB (National Plant Board) (1999) Safeguarding American Plant Resources, a Stakeholder Review of the APHIS-PPQ Safeguarding System. Available at: http://www.nationalplantboard.org/policy/safeguard.html (accessed 25 June 2007).

NRC (National Research Council) (1993) Pesticides in the Diets of Infants and Children. Available at: http://www.uwmc.uwc.edu/geography/350/pesticides_in_the_diets_of_infan.htm (accessed 25 June 2007).

Obenland, D.M., Jang, E.B., Aung, L.H. and Zettler, L. (1998) Tolerance of lemons and the Mediterranean fruit fly to carbonyl sulfide quarantine fumigation. *Crop Protection* 17, 219–224.

Page, B.B.C., Bendall, M.J., Carpenter, A. and Epenhuijsen, C.W. van (2002) Carbon dioxide fumigation of *Thrips tabaci* in export onions. *New Zealand Plant Protection* 55, 303–307.

Park, M.-G., Heo, J.-Y., Jeong, W.-C., Choi, D.-H. and Lee, K.-S. (2006) HCN with forced aeration in imported oranges. In: Anon. (ed.) *Proceedings of 2006 International Research Conference on Methyl Bromide Alternatives and Emission Reductions* 145, 1–4. Available at: www.mbao.org (accessed 25 June 2007).

Parsell, D.A., Taulien, J. and Lindquist, S. (1993) The role of heat shock proteins in thermotolerance. In: Ellis, R.J., Laskey, R.A. and Lorimer, G.H. (eds) *Molecular Chaperones*. Chapman and Hall, London, UK.

Paull, R.E. and Armstrong, J.W. (eds) (1994) *Insect Pests and Fresh Horticultural Products: Treatments and Responses*. CAB International, Wallingford, UK.

Paull, R.E. and McDonald, R.E. (1994) Heat and cold treatments. In: Paull, R.E. and Armstrong, J.W. (eds) *Insect Pests and Fresh Horticultural Products: Treatments and Responses*. CAB International, Wallingford, UK, pp. 191–222.

Penteado, A.L., Eblen, B.S. and Miller, A.J. (2004) Evidence of *Salmonella* internalization into fresh mangoes during simulated postharvest insect disinfestation procedures. *Journal of Food Protection* 67, 181–184.

Pheloung, P. and Macbeth, F. (2002) Export inspection: adding value to Australia's grain. In: Wright, E.J., Banks, H.J. and Highley, E. (eds) *Stored Grain in Australia*. Proceedings of the Australian Postharvest Technical Conference 2000, Adelaide 1–4 August 2000. Commonwealth Scientific and Industrial Research Organization (CSIRO), Stored Grain Research Laboratory, Canberra, Australian Capital Territory, Australia, pp. 15–17.

Phillips, D. (1986) Irradiated fruit gets Miami okay. *The Packer* 93, 1A and 6A.

Pielke, R.A. Jr and Oreskes, N. (2005) Consensus about climate change? *Science* 308, 952–954.

Pimentel, D., Lach, L., Zuniga, R. and Morrison, D. (2002) Environmental and economic costs of alien arthropods and other organisms in the USA: In: Hallman, G.J. and Schwalbe, C.P. (eds) *Invasive Arthropods in Agriculture: Problems and Solutions*. Science Publishers, Enfield, New Hampshire, pp. 107–117.

Pimentel, D., Pimentel, M. and Wilson, A. (2007) Plant, animal, and microbe invasive species in the USA and the world. In: Nentwig, W. (ed.) *Biological Invasions*. Springer, Berlin, pp. 316–330.

Pinniger, D.B. (1990) Sampling and trapping insect populations, the importance of environment, insects and trade. In: Fleurat-Lessard, F. and Ducom, P. (eds) *Proceedings of the Fifth International Working Conference on Stored-Product Protection*. Bordeaux, France, pp. 1297–1306.

Potenza, M.R., Yasuo-ka, S.T., Giordano, R.B.P. and Raga, A. (1989) Irradiação de frutos de laranja infestados com larvas da mosca das frutas *Ceratitis capitata* (Weid., 1824) [Abstract]. *Congresso Brasileiro de Entomologia*, 1989, Belo Horizonte, Brazil.

Potter, M.A., Carpenter, A. and Stocker, A. (1990) Response surfaces for controlled atmosphere and temperature by species: implications for disinfestation of fresh produce for export. In: Beattie, B.B. (ed.) *Managing Postharvest Horticulture in Australasia*. Occasional Publication No. 46. Australian Institute of Agricultural Science, Melbourne, Australia, pp. 183–190.

Potter, M.A., Carpenter, A., Stocker, A. and Wright, S. (1994) Controlled atmospheres for the postharvest disinfestation of *Thrips obscuratus* (Thysanoptera: Thripidae). *Journal of Economic Entomology* 87, 1251–1255.

Powell, M.E. (2003) Modeling the response of the Mediterranean fruit fly (Diptera: Tephritidae) to cold treatment. *Journal of Economic Entomology* 96, 300–310.

Prange, R.K. and Lidster, P.D. (1992) Controlled-atmosphere effects on blueberry maggot and lowbrush blueberry fruit. *HortScience* 27, 1094–1096.

Priesler, H.K. and Robertson, J.L. (1992) Estimation of treatment efficacy when numbers of test subjects are unknown. *Journal of Economic Entomology* 85, 1033–1040.

Raga, A. (1990) Uso da Radiação Gama na Desinfestação de Mangas Destinadas à Exportação em Relação à *Ceratitis capitata* (Wied., 1824), *Anastrepha fraterculus* (Weid., 1830) e *Anastrepha obliqua* (Macquart, 1835) (Diptera, Tephritidae). MSc thesis, University of São Paulo, Brazil.

Raga, A. (1996) Incidência de Mosca-das-Frutas em Café e Citros e Tratamento Quarentenário de Frutos Cítricos com Radiação Gama. PhD dissertation, University of São Paulo, Brazil.

Rahman, R., Rigney, C. and Busch-Petersen, E. (1990) Irradiation as a quarantine treatment against *Ceratitis capitata* (Diptera: Tephritidae): anatomical and cytogenetic changes in mature larvae after gamma irradiation. *Journal of Economic Entomology* 83, 1449–1454.

Rahman, R., Bhuiya, A.D., Huda, S.M.S., Shahjahan, R.M., Nahar, G. and Wadud, M.A. (1992) Anatomical changes in the mature larvae of two *Dacus* spp. following irradiation. In: International Atomic Energy Agency (IAEA) (ed.) *Use of Irradiation as a Quarantine Treatment of Food and Agricultural Commodities.* IAEA, Vienna, pp. 133–139.

Richardson, H.H. (1952) Cold treatment of fruits. In: *Insects: the Yearbook of Agriculture.* US Department of Agriculture (USDA), Washington, DC, pp. 404–406.

Riherd, C., Nguyen, R. and Brazzel, J.R. (1994) Pest free areas. In: Sharp, J.L. and Hallman, G.J. (eds) *Quarantine Treatments for Pests of Food Plants.* Westview Press, Boulder, Colorado, pp. 213–223.

Robacker, D.C. and Landolt, P.J. (2002) Importance and use of attractants. In: Hallman, G.J. and Schwalbe, C.P. (eds) *Invasive Arthropods in Agriculture: Problems and Solutions.* Science Publishers, Enfield, New Hampshire, pp. 169–205.

Roberts, D. and Krissoff, B. (2004) Regulatory Barriers in International Horticultural Markets. Outlook. USDA Economic Research Service. Electronic Report. Available at: http://www.ers.usda.gov/ (accessed 25 May 2007).

Robertson, J.L., Smith, K.C., Savin, N.E. and Lavigne, R.J. (1984) Effects of dose selection and sample size on the precision of lethal dose estimates in dose-mortality regression. *Journal of Economic Entomology* 77, 833–837.

Robertson, J.L., Preisler, H.K., Frampton, E.R. and Armstrong, J.W. (1994) Statistical analyses to estimate efficacy of disinfestation treatments. In: Sharp, J.L. and Hallman, G.J. (eds) *Quarantine Treatments for Pests of Food Plants.* Westview Press, Boulder, Colorado, pp. 47–65.

Rohwer, G.C. and Williamson, D.L. (1983) Pest risk evaluation in regulatory entomology. *Bulletin of the Entomological Society of America* Winter, pp. 41–46.

Ruan, R., Ye, X., Chen, P., Doona, C. and Yang, T. (2004) Developments in ohmic heating. In: Richardson, P. (ed.) *Improving the Thermal Processing of Food.* Woodhead, Cambridge, UK, pp. 224–252.

Ryan, R. and Bishop, S. (2003) Vapormate™: non-flammable ethyl formate/liquid carbon dioxide fumigant mixture. In: Wright, E.J., Webb, M.C. and Highley, E. (eds) *Stored Grain in Australia 2003.* CSIRO (Commonwealth Scientific and Industrial Research Organization), Canberra, Australian Capital Territory, Australia, pp. 190–192.

Saeed, S., Kwon, Y.-J. and Kausar, T. (2006) Irradiation control of *Plodia interpunctella* L. (Lepidoptera: Pyralidae) in dehydrated ginseng (*Panax ginseng*). *Pakistan Journal of Zoology* 38, 33–37.

Saul, S. and Seifert, J. (1990) Methoprene on papayas: persistence and toxicity to different development stages of fruit flies (Diptera: Tephritidae). *Journal of Economic Entomology* 83, 901–904.

Saul, S.H., Mau, R.F.L. and Oi, D. (1985) Laboratory trials of methoprene-impregnated waxes for disinfesting papayas and peaches of the Mediterranean fruit fly (Diptera: Tephritidae). *Journal of Economic Entomology* 78, 652–655.

Saul, S.H., Mau, R.F.L., Kobayashi, R.M., Tsuda, D.M. and Nishina, M.S. (1987) Laboratory trials of methoprene-impregnated waxes for preventing survival of adult oriental fruit flies (Diptera: Tephritidae) from infested papayas. *Journal of Economic Entomology* 80, 494–496.

Saunders, G.W. and Elder, R.J. (1966) Sterilisation of banana fruit infested with banana fruit fly. *Queensland Journal of Agricultural Science* 23, 81–85.

Scruton, T. (2000) Inspectors find larvae in Mexican mangoes. *The Packer*, 20 March, p. A7.

Seo, S.T. and Tang, C.-S. (1982) Hawaiian fruit flies: (Diptera: Tephritidae): toxicity of benzyl isothiocyanate against eggs or 1st instars of three species. *Journal of Economic Entomology* 75, 1125–1132.

Seo, S.T., Chambers, D.L., Akamine, E.K., Komura, M. and Lee, C.Y.L. (1972) Hot water-ethylene dibromide fumigation-refrigeration treatment for mangoes infested by oriental and Mediterranean fruit flies. *Journal of Economic Entomology* 65, 1372–1374.

Seo, S.T., Kobayashi, R.M., Chambers, D.L., Dollar, D.M. and Hanaoka, M. (1973) Hawaiian fruit flies in papaya, bell pepper, and eggplant: quarantine treatment with gamma irradiation. *Journal of Economic Entomology* 66, 937–939.

Seo, S.T., Tang, C.-S., Sanidad, S. and Takenaka, T.H. (1983) Hawaiian fruit flies (Diptera: Tephritidae): variation of index of infestation with benzyl isothiocyanate concentration and color of maturing papayas. *Journal of Economic Entomology* 76, 535–538.

Sequeira, R.A. (2002) Agricultural risk assessment. In: Hallman, G.J. and Schwalbe, C.P. (eds) *Invasive Arthropods in Agriculture: Problems and Solutions*. Science Publishers, Enfield, New Hampshire, pp. 140–156.

Sharp, J.L. (1988) Status of hot water immersion quarantine treatment for Tephritidae immatures in mangoes. *Proceedings of the Florida State Horticultural Society* 101, 195–197.

Sharp, J.L. (1994) Hot water immersion. In: Sharp, J.L. and Hallman, G.J. (eds) *Quarantine Treatments for Pests of Food Plants.* Westview Press, Boulder, Colorado, USA, pp. 133–170.

Sharp, J.L. and Heather, N.W. (2002) Quarantine treatments for pests of tropical fruits. In: Peña, J.E., Sharp, J.L. and Wysoki, M. (eds) *Tropical Fruit Pests and Pollinators.* CAB International, Wallingford, UK, pp. 391–406.

Sharp, J.L. and Picho-Martinez, H. (1990) Hot-water treatment to control fruit flies in mangoes imported into the USA from Peru. *Journal of Economic Entomology* 83, 1940–1943.

Shellie, K.C. and Mangan, R.L. (2000) Postharvest disinfestation heat treatments: response of fruit and fruit fly larvae to different heating media. *Postharvest Biology and Technology* 21, 51–60.

Shellie, K.C. and Mangan, R.L. (2002) Hot water immersion as a quarantine treatment for large mangoes: artificial *versus* cage infestation. *Journal of the American Society of Horticultural Science* 127, 430–434.

Shellie, K.C., Mangan, R.L. and Ingle, S.J. (1997) Tolerance of grapefruit and Mexican fruit fly larvae to heated controlled atmospheres. *Postharvest Biology and Technology* 10, 179–186.

Shultz, R.H. Jr (1999) *The Secret War Against Hanoi: the Untold Story of Spies, Saboteurs, and Covert Warriors in North Vietnam.* HarperCollins, New York.

Siddiqui, K.H. and Dey, D.D. (2002) *Artificial Diets – a Tool for Insect Mass Rearing.* Jyoti Publishers, Inderpuri, India.

Singh, P. and Moore, R.F. (1985) *Handbook of Insect Rearing.* Vols. I and II. Elsevier, Amsterdam.

Smith, E.S.C. (1977) Studies on the biology and commodity control of the banana fruit fly, *Dacus musae* (Tryon) in Papua, New Guinea. *Papua, New Guinea Agricultural Journal* 28, 47–56.

Smoot, J.J. and Segall, R.H. (1963) Hot water as a postharvest control of mango anthracnose. *Plant Disease Reporter* 47, 739–742.

Snelson, J.T. (1987) *Grain Protectants*. Australian Centre for International Agricultural Research, Canberra, Australia.

Soderstrom, E.L., Brandl, D.G. and Mackey, B. (1991) Responses of *Cydia pomonella* (L.) (Lepidoptera: Tortricidae) adults and eggs to oxygen deficient or carbon dioxide enriched atmospheres. *Journal of Stored Products Research* 27, 95–101.

Soderstrom, E.L., Brandl, D.G. and Mackey, B.E. (1993) High temperature for control of *Asynonychus godmani* (Coleoptera: Curculionidae) eggs on lemon fruit. *Journal of Economic Entomology* 86, 1773–1780.

Soderstrom, E.L., Brandl, D.G. and Mackey, B.E. (1996) High temperature alone and combined with controlled atmospheres for control of diapausing codling moth (Lepidoptera: Tortricidae) in walnuts. *Journal of Economic Entomology* 89, 144–147.

Soma, Y., Matsuoka, Y., Naito, H., Tsuchiya, Y. Misumi, T. and Kawakami, F. (2002) *Research Bulletin of the Plant Protection Service, Japan* 38, 9–12.

Southwood, T.R.E. and Henderson, P.A. (2000) *Ecological Methods*, 3rd edn. Blackwell Science, Oxford, UK.

Stark, J.D. (1994) Chemical fumigants. In: Paull, R.E. and Armstrong, J.W. (eds) *Insect Pests and Fresh Horticultural Products: Treatments and Responses*. CAB International, Wallingford, UK, pp. 69–84.

Stewart, W.E., Becker, B.R., Greer, M.E. and Stickler, L.A. (1990) An experimental method of approximating effective heat transfer coefficients in food products. *Transactions of the American Society of Heating, Refrigeration and Air-Conditioning Engineers* 96, 142–147.

Suckling, D.M., Hackett, J.K., Chhagan, A., Barrington, A. and El-Sayed, A.M. (2006) Effect of irradiation on female painted apple moth *Teia anartoides* (Lep., Lymantriidae) sterility and attractiveness to males. *Journal of Applied Entomology* 130, 167–170.

Sugimoto, T., Furasawa, K. and Mizobuchi, M. (1983) The effectiveness of vapour heat treatment against the Oriental fruit fly, *Dacus dorsalis* Hendel in green peppers and fruit tolerance to treatment (in Japanese). *Research Bulletin of the Plant Protection Service of Japan* 19, 81–88.

Sunagawa, K., Kume, K., Ishikawa, A., Sugimoto, T. and Tanabe, K. (1988) Efficiency of vapour heat treatment for bitter momordica fruit infested with melon fly, *Dacus cucurbitae* (Coquillett). *Research Bulletin of the Plant Protection Service of Japan* 24, 1–5.

Sutherst, R.W. and Maywald, G.F. (1985) A computerised system for matching climates in ecology. *Agriculture Ecosystems and Environment* 13, 281–299.

Swaine, G., Hargreaves, P.A., Jackson, D.E. and Corcoran, R.J. (1984a) Dimethoate dipping of tomatoes against Queensland fruit fly *Dacus tryoni* (Froggatt). *Australian Journal of Experimental Agriculture and Animal Husbandry* 24, 447–449.

Swaine, G., Melksham, K.J. and Corcoran, R.J. (1984b) Dimethoate dipping of Kensington mango against Queensland fruit fly. *Australian Journal of Experimental Agriculture and Animal Husbandry* 14, 620–623.

Szczepanik, M. and Ignatowicz, S. (1994) Identification of irradiated insects: changes in the midgut of the confused flour beetle, *Tribolium confusum* Duval., induced by gamma radiation. *Roczn. Nauk Roin., Seria E.* 24, 69–77.

Tang, J., Mitcham, E., Wang, S. and Lurie, S. (eds) (2007) *Heat Treatments for Postharvest Pest Control: Theory and Practice*. CAB International, Wallingford, UK.

Thomas, D.B. and Mangan, R.L. (1995) Morbidity of the pupal stage of the Mexican and West Indian fruit flies (Diptera: Tephritidae) induced by hot-water immersion in the larval stage. *Florida Entomologist* 78, 235–246.

Thomas, M.C., Heppner, J.B., Woodruff, R.E., Weems, H.V., Steck, G.J. and Fasulo, T.R. (2005) Featured Creatures: Mediterranean fruit fly *Ceratitis capitata*. University of Florida Institute of Food and Agricultural Services and Florida Department of Agriculture and Consumer Services web site. Available at: http://creatures.ifas. ufl.edu/fruit/Mediterranean_fruit_fly.htm (accessed 25 June 2007).

Thompson, A. (1990) Efecto de cera y aceite en frutos de cherimoya (*Annona squamosa* Mill.) cv. Concha Lisa, sobre el control de *Brevipalpus chilensis* Baker y evaluación del comportamiento de la fruta en pos-cosecha. Thesis in agronomy, Universidad Católica de Valparaiso, Chile.

Thompson, A.K. (ed.) (1998) *Controlled Atmosphere Storage of Fruits and Vegetables.* CAB International, Wallingford, UK.

Toba, H.H. and Moffitt, H.R. (1991) Controlled-atmosphere cold storage as a quarantine treatment for nondiapausing codling moth (Lepidoptera: Tortricidae) larvae in apples. *Journal of Economic Entomology* 84, 1316–1319.

Torres-Rivera, Z. and Hallman, G.J. (2007) Low-dose ionizing irradiation phytosanitary treatment against Mediterranean fruit fly (Diptera: Tephritidae) in fruit. *Florida Entomologist* 90, 343–346.

Tuncbilek, A.S. (1997) Susceptibility of the saw-toothed grain beetle, *Oryzaephilus surinamensis* (L.), to gamma radiation. *Journal of Stored Product Research* 33, 331–334.

Unahawutti, U., Poomthong, M., Intarakumheng, R., Worawiisitthumrong, W., Lapasathukool, C., Smitasiri, E., Srisook, P. and Ratanawaraha, C. (1992) Vapor heat as plant quarantine treatment of 'Nsg. Klarngwan', 'Nam Dorkmai', 'Rad' and 'Pimsen Daeng' mangoes infested with fruit flies (Diptera: Tephritidae). Department of Agriculture Technical Report, Chatuchak, Thailand.

Unduragga, M.F. and Lopez, L.E. (1992) Use of wax and soapy water, alone or combined as a quarantine method of control of *Brevipalpus chilensis* Baker on cherimoyas (*Annona squamosa* Mill.) (Report). Universidad Católica de Valparaiso, Valparaiso, Chile.

UNIDO (United Nations Industrial Development Organisation) (2007) The Montreal Protocol – a Success Story. Available at: http://www.unido.org/en/doc/50444 (accessed 25 June 2007).

Unterstenhofer, G. (1960) Lebaycid, a new insecticide of low toxicity. *Hofchen-Briefe* 13, 44–52.

USGAO (US Government Accountability Office) (2006) Homeland Security: Management and Coordination Problems Increase the Vulnerability of US Agriculture to Foreign Pests and Diseases. Available at: http://www.gao.gov/ new.items/d06644.pdf (accessed 25 June 2007).

USGS (US Geological Survey) (2005) *Geographic Information Systems.* US Geological Survey Poster. Available at: http://erg.usgs.gov/isb/pubs/gis_poster/ (accessed 25 June 2007).

Vail, P.V., Tebbets, J.S. and Mackey, B.E. (1993a) Quarantine treatments: a biological approach to decision making for selected hosts of codling moth (Lepidoptera: Tortricidae). *Journal of Economic Entomology* 86, 70–75.

Vail, P.V., Tebbets, J.S., Curtis, C.E. and Jenner, K.E. (1993b) Infestation rates and biological observations after harvest of codling moth (Lepidoptera: Tortricidae) in stored in-shell walnuts. *Journal of Economic Entomology* 86, 1761–1765.

Vera, M.T., Rodriguez, R., Segura, D.F., Cladera, J.L. and Sutherst, R.W. (2002) Potential geographical distribution of the Mediterranean fruit fly, *Ceratitis capitata* (Diptera: Tephritidae), with emphasis on Argentina and Australia. *Environmental Entomology* 31, 1009–1022.

Wadley, F.M. (1949) Dosage-mortality correlation with number treated estimated from a parallel sample. *Annals of Applied Biology* 36, 196–202.

Walker, G.P., Morse, J.G. and Arpaia, M.L. (1996) Evaluation of a high-pressure washer for postharvest removal of California red scale (Homoptera: Diaspididae) from citrus fruit. *Journal of Economic Entomology* 89, 148–155.

Wang, S., Tang, J., Johnson, J.A., Mitcham, E.J., Hansen, J.D., Hallman, G.J., Drake, S.R. and Wang, Y. (2003) Dielectric properties of fruits and insect pests as related to radio frequency and microwave treatments. *Biosystems Engineering* 85, 201–212.

Wang, S., Tang, J., Sun, T., Mitcham, E.J., Koral, T. and Birla, S.L. (2006) Considerations in design of commercial radio frequency treatments for postharvest pest control in in-shell walnuts. *Journal of Food Engineering* 77, 304–312.

Wheelis, M. (2002) Biotechnology and biochemical weapons. *Nonproliferation Review* 9, 48–53.

Wheelis, M. (2004) A short history of biological warfare and weapons: In: Chevrier, M.I., Chomiczewski, K., Garrigue, H., Granasztói, G., Dando, M.R. and Pearson, G.S. (eds) *The Implementation of Legally Binding Measures to Strengthen the Biological and Toxin Weapons Convention*. Springer, New York, pp. 15–32.

Wheelis, M., Casagrande, R. and Madden, L.V. (2002) Biological attack on agriculture: low-tech, high-impact bioterrorism. *BioScience* 52, 569–576.

White, G. and Blackett, D. (1988) Grain storage insect control recommendations. *Farm Note*. Queensland Department of Primary Industries, Brisbane, Australia.

White, G.G. and Lambkin, T.A. (1990) Baseline responses to phosphine and resistance status of stored-grain beetle pests in Queensland, Australia. *Journal of Economic Entomology* 83, 1738–1744.

Whiting, D.C. and Heuvel, J. van den (1995) Oxygen, carbon dioxide, and temperature effects on mortality responses of diapausing *Tetranychus urticae* (Acari: Tetranychidae). *Journal of Economic Entomology* 88, 331–336.

Whiting, D.C. and Hoy, L.E. (1997) High-temperature controlled atmosphere and air treatments to control obscure mealy bug (Hemiptera: Pseudococcidae) on apples. *Journal of Economic Entomology* 90, 546–550.

Whiting, D.C. and Hoy, L.E. (1998) Effect of temperature establishment time on the mortality of *Epiphyas postvittana* (Lepidoptera: Tortricidae) larvae exposed to a high-temperature controlled atmosphere. *Journal of Economic Entomology* 91, 287–292.

Whiting, D.C., Foster, S.P. and Maindonald, J.H. (1991) Effects of oxygen, carbon dioxide and temperature on the mortality responses of *Epiphyas postvittana* (Lepidoptera: Tortricidae). *Journal of Economic Entomology* 84, 1544–1549.

Whiting, D.C., Foster, S.P., Heuvel, J. van den and Maindonald, J.H. (1992a) Comparative mortality responses of four tortricid (Lepidoptera) species to a low oxygen-controlled atmosphere. *Journal of Economic Entomology* 85, 2305–2309.

Whiting, D.C., Heuvel, J. van den and Foster, S.P. (1992b) Potential of low oxygen/moderate carbon dioxide atmospheres for postharvest disinfestation of New Zealand apples. *New Zealand Journal of Crop and Horticultural Science* 20, 217–222.

Whiting, D.C., O'Connor, G.M., Heuvel, J. van den and Maindonald, J.H. (1995) Comparative mortalities of six tortricid (Lepidoptera) species to two high-temperature controlled atmospheres and air. *Journal of Economic Entomology* 88, 1365–1370.

Whiting, D.C., Hoy, L.E., Connolly, P.G. and McDonald, R.M. (1998a) Effects of high-pressure water jets on armoured scale insects and other contaminants of harvested kiwi fruit. *Proceedings of the 51st New Zealand Plant Protection Conference 1998*, pp. 211–215.

Whiting, D.C., Hoy, L.E., Maindonald, J.H., Connolly, P.G. and McDonald, R.M. (1998b) High-pressure washing treatments to remove obscure mealy bug (Homoptera: Pseudococcidae) and lightbrown apple moth (Lepidoptera: Tortricidae) from harvested apples. *Journal of Economic Entomology* 91, 1458–1463.

WHO (World Health Organization) (1999) *High-dose Irradiation: Wholesomeness of Food Irradiated with Doses above 10 kGy.* Report of a Joint Food and Agriculture Organization (FAO)/International Atomic Energy Agency (IAEA)/WHO study group 15–20 September 1997. WHO Technical Report 890. WHO, Geneva, Switzerland.

WHO (World Health Organization) (2007) *IARC Monographs on the Evaluation of Carcinogenic Risks to Humans.* WHO. Available at: http://monographs.iarc.fr/ ENG/Monographs/ (accessed 25 June 2007).

Williams, P. and Muhunthan, M. (1998) Fumigants for postharvest control of insect pests of cut flowers. *Acta Horticulturae* 464, 291–296.

Williams, P., Hepworth, G., Goubran, F., Muhunthan, M. and Dunn, K. (2000) Phosphine as a replacement for methyl bromide for postharvest disinfestation of citrus. *Postharvest Biology and Technology* 19, 193–199.

Williams, P., Green, K.M. and Swanson, B. (2005) A cold treatment for postharvest control of western flower thrips on strawberry runners. *Australian Journal of Experimental Agriculture* 45, 1649–1651.

Williamson, D.L., Summy, K.R., Hart, W.G., Sanchez, R.M., Wolfenbarger, D.A. and Bruton, B.D. (1986) Efficacy and phytotoxicity of methyl bromide as a fumigant for the Mexican fruit fly (Diptera: Tephritidae) in grapefruit. *Journal of Economic Entomology* 79, 172–175.

Williamson, M.R. and Winkelman, P.M. (1989) Commercial scale heat treatment for disinfestation of papaya. *American Society of Engineers Paper* 89, 6054.

Willis, J.C. (1966) *A Dictionary of the Flowering Plants and Ferns.* Cambridge University Press, Cambridge, UK.

Winks, R.G. (1983) Laboratory culturing of some storage pests. In: Champ, B.R. and Highley, E. (eds) *Proceedings of the Australian Development Assistance Course on the Preservation of Stored Cereals.* Vol. 1. CSIRO (Commonwealth Scientific and Industrial Research Organization), Canberra, Australian Capital Territory, Australia, pp. 219–239.

Winks, R.G. (1984) Fumigants and their application. In: Champ, B.R. and Highley, E. (eds) *Proceedings of the Australian Development Assistance Course on the Preservation of Stored Cereals.* Vol. 2. CSIRO (Commonwealth Scientific and Industrial Research Organization), Canberra, Australian Capital Territory, Australia, pp. 721–730.

Winks, R.G. (1993) The development of SIROFLO in Australia. In: Navarro, S. and Donahaye, E. (eds) *Proceedings of the International Conference on Controlled Atmospheres and Fumigation in Grain Storages, Winnipeg, Canada, 1992.* Caspit Press, Jerusalem, Israel, pp. 399–410.

Worner, S.P. (1994) Predicting the establishment of exotic pests in relation to climate. In: Sharp, J.L. and Hallman, G.J. (eds) *Quarantine Treatments for Pests of Food Plants.* Westview Press, Boulder, Colorado, pp. 11–32.

Worner, S.P. (2002) Predicting the invasive potential of exotic insects. In: Hallman, G.J. and Schwalbe, C.P. (eds) *Invasive Arthropods in Agriculture: Problems and Solutions.* Science Publishers, Enfield, New Hampshire, pp. 119–137.

Worner, S.P. and Gevrey, M. (2006) Modelling global insect pest species assemblages to determine risk of invasion. *Journal of Applied Ecology* 43, 858–867.

Wright, E.J., Ren, Y.L. and Dowsett, H. (2002) Cyanogen: a new fumigant with potential for timber. In: Anon. (ed.) *Proceedings of 2002 International Research Conference on Methyl Bromide Alternatives and Emission Reductions.* Available at: www.mbao.org (accessed 25 June 2007).

WTO (World Trade Organization) (2007a) The WTO Agreement on the Application of Sanitary and Phytosanitary Measures (SPS Agreement). Available at: http://www.wto.org/english/tratop_e/sps_e/spsagr_e.htm (accessed 25 June 2007).

WTO (World Trade Organization) (2007b) International Trade Statistics (World Trade Organization Statistics Database web site). Available at: http://www.wto.org/english/res_e/statis_e/statis_e.htm (accessed 25 June 2007).

WTO (World Trade Organization) (2007c) The Variety Testing Case. Available at: http://www.wto.org/english/tratop_e/sps_e/sps_agreement_cbt_e/c5s5p1_e.htm (accessed 25 June 2007).

Yalemar, J.A., Hara, A.H., Saul, S.H., Jang, E.B. and Moy, J.H. (2001) Effects of gamma irradiation on the life stages of yellow flower thrips, *Frankliniella schultzei* (Trybom) (Thysanoptera: Thripidae). *Annals of Applied Biology* 138, 263–268.

Yin, X., Wang, S., Tang, J. and Hansen, J.D. (2006) Thermal resistance of fifth-instar *Cydia pomonella* (L.) (Lepidoptera: Tortricidae) as affected by pre-treatment conditioning. *Journal of Stored Product Research* 42, 75–85.

Yokoyama, V.Y. (1994) Fumigation. In: Sharp, J.L. and Hallman, G.J. (eds) *Quarantine Treatments for Pests of Food Plants*. Westview Press, Boulder, Colorado, pp. 67–87.

Yokoyama, V.Y. and Miller, G.T. (2002) Bale compression and hydrogen phosphide fumigation to control cereal leaf beetle (Coleoptera: Chrysomelidae) in exported rye straw. *Journal of Economic Entomology* 95, 513–519.

Yokoyama, V.Y. and Miller, G.T. (2003) Quarantine control of Hessian fly (Diptera: Cecidomyiidae) in exported hay: a new treatment for large-size, polypropylene fabric-wrapped bales and a 3-d fumigation for compressed standard bales. *Journal of Economic Entomology* 96, 1340–1344.

Yonow, T., Zalucki, M.P., Sutherst, R.W., Dominiak, B.C., Maywald, G.F., Maelzer, D.A. and Kritikos, D.J. (2004) Modelling the population dynamics of the Queensland fruit fly, *Bactrocera (Dacus) tryoni*. A cohort based approach incorporating the effects of weather. *Ecological Modelling* 173, 9–30.

Zachariassen, K.E. (1985) Physiology of cold tolerance in insects. *Physiological Reviews* 65, 799–832.

Zakladnoi, G.A., Pertzovski, E., Menshinin, A. and Cherepkov, B. (1981) Mukomol'no elevat. Kombi. *Promyshl* 6, 29.

Zee, F.T., Nishina, M.S., Chan, H.T. and Nishijima, K.A. (1989) Blossom end defects and fruit infestation in papayas following hot water quarantine treatment. *HortScience* 24, 323–325.

Zettler, J.L. and Cuperis, G.W. (1990) Pesticide resistance in *Tribolium castaneum* (Coleoptera: Tenebrionidae) and *Rhyzopertha dominica* (Coleoptera: Bostrychidae) in wheat. *Journal of Economic Entomology* 83, 1677–1681.

Zettler, J.L. and Leesch, J.G. (2000) Sulfuryl fluoride: a disinfestation treatment for walnuts and almonds. *Methyl Bromide Alternatives* 2, 5–6.

Zolfagharieh, H. (2004) Irradiation to control *Plodia interpunctella* and *Oryzaphilus surinamensis* in pistacios and dates. *Irradiation as a Phytosanitary Treatment of Food and Agricultural Commodities*. IAEA-TECDOC-1427. International Atomic Energy Agency, Vienna, pp. 101–109.

Index